# Decoupling and Layout of Digital Printed Circuits

R. Kenneth Keenan, Ph.D.

TKC • THE KEENAN CORPORATION

Consulting Electronics Engineers

The Keenan Building 8609 66th Street North Pinellas Park, FL 34666
TELEX: 469030 FAX: (813) 544-2596 TELE: (813) 544-2594

Second Printing, September 1987

Library of Congress Card Catalog Number 85-72123

To What's-her-face's "Tara",

JELEPHANT

ACKNOWLEDGEMENTS

Prior to publication, this book was reviewed by Messrs. Frank Justice and Dave Terrell of TKC, and by Dr. Scott Bennett of Hewlett-Packard. All have made substantial contributions to the clarity of presentation. Mr. Dave Terrell did the software and plotting work for impedance of PCB traces (Appendix 1), and Dr. Scott Bennett provided the Index.

## TABLE OF CONTENTS

TABLE OF CONTENTS (Cont'd)

TABLE OF CONTENTS (Cont'd)

TABLE OF CONTENTS (CONCLUDED)

# LIST OF FIGURES

## LIST OF FIGURES (Cont'd)

## LIST OF FIGURES (CONCLUDED)

# LIST OF TABLES

# 1.0 INTRODUCTION

## 1.1 Importance of Decoupling

By "decoupling" is meant provision of means to dissociate circuit functions from the power bus serving that circuit. In practice, "means" include decoupling and power-entry capacitors, ferrite beads, and appropriate printed-circuit board (PCB) layout. All of these subjects are addressed in this book. Both emissions and susceptibility aspects of decoupling are considered.

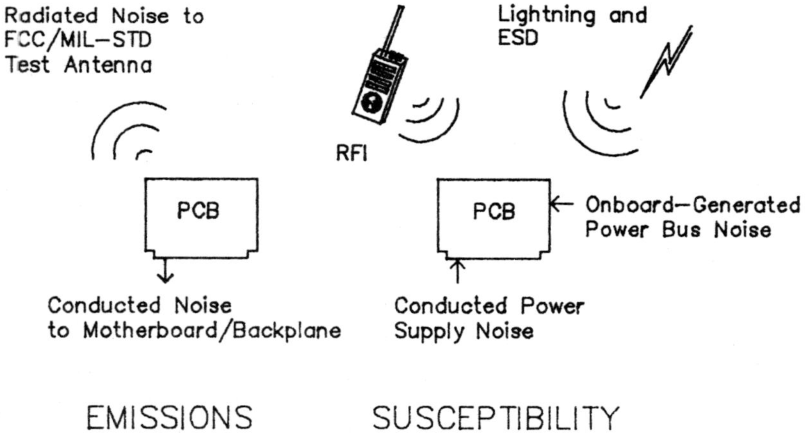


Figure 1.0-1: EMISSIONS AND SUSCEPTIBILITY ASPECTS OF DECOUPLING

Poor decoupling increases radiated and conducted emissions* from equipment and, also, increases susceptibility of equipment to both internally and externally-generated noise. For example, erratic and/or intermittent digital circuit functions due to power bus noise constitute a susceptibility to conducted noise. Should that noise be caused by operations of other digital circuits connected to the power bus, the susceptibility is said to arise from internally-generated noise. Malfunctions due to power bus variations

* **Radiated emissions** are those received by an antenna. **Conducted emissions** are usually measured on the power cord supplying the equipment, but the description also applies to noise on an internal power bus.

induced by a hand-held radio transceiver constitute a susceptibility to externally-generated radio-frequency interference (RFI).

Early digital logic decoupling concerns were largely restricted to maintaining a small enough level of power bus noise to remain within the noise immunity limits of a given logic family. The singularity of this concern permitted relatively few decoupling capacitors and relatively poor PCB layouts. For commercial equipment marketed in the U.S., FCC specifications restricting emissions went into full effect in 1983. Emissions considerations mandate much better decoupling practices than those soley for the purposes of reducing susceptibility to power bus noise. The costs of emissions retrofits suggest that a more economical course of action is to pay close attention to decoupling during initial design.

There are, of course, issues other than decoupling affected by emissions limits, but experience suggests that the "superstructure" of the PCB - decoupling - be in order before addressing less mundane considerations.

In military equipment, at least some of the "necessity" for extensive shielding can be alleviated by paying additional attention to decoupling.

1.2 Purpose and Motivation

**The purpose of this book is to provide design guidelines for decoupling digital logic cicuits arrayed (layed out) on PCBs.** These guidelines permit decoupling dual inline packages (DIPs) and other integrated circuit packages from one another in ways which minimize susceptibility to noise. Concurrently, they facilitate compliance with radiated and conducted emissions specifications applicable to the equipment in which the PCB is used.

Typical commercial specifications include FCC Part 15, Subpart J ("Computing Device{s}") and its European counterparts (e.g., VDE 0871) limiting radiated and conducted emissions from computer-based equipment. Typical military specifications include MIL-STD-461B ("Electromagnetic Emission and Susceptibility Requirements for the Control of Electromagnetic Interference") and TEMPEST (emissions only).

This work is motivated by consulting and seminar experiences at TKC, during which it has become clear that many designers accept, without question, decoupling recommendations of DIP and/or capacitor manufacturers. The attendant lack of analytical insight can be costly and lead to poor designs when those recommendations are extrapolated beyond the usually narrow frameworks in which they are made. However, there is little in the present literature which comprehensively deals with susceptibility and emissions issues implicit in decoupling. This work is an attempt to fill that void.

## 1.3 Scope and Emphases

Military and commercial multilayer PCBs are within the scope of the analysis presented here, but emphasis is on two-sided PCBs, where some decoupling problems are exacerbated by the absence of a ground plane. The two-sided PCB is the more difficult problem, and it is of great economic importance.

Insofar as costs are considered important, this book is biased in the direction of commercial, as opposed to military, systems. The economic implications of selection of decoupling capacitors are substantial: a typical microprocessor-based product employs at least many tens and perhaps many hundreds of decoupling capacitors. The cost of those capacitors are considered in development of the material in this book.

It will be shown in Section 3.8 that, in most designs, those radiated emissions due to PCB trace loop antenna effects are primarily dependent on layout of clock traces, rather than decoupling considerations. Clock trace layout issues are not addressed in this book, but are addressed in the author's previous book [1]. As discussed in Section 3.8, decoupling can have a strong effect on equipment radiation via means other than loop-based radiation, and, for that reason, is important in minimizing equipment radiation.

## 1.4 Organization

"Rules of thumb" applicable to decoupling in the majority of cases encountered in practice are summarized in Section 2.0 and are developed in more detail in Section 4.0. Both of those Sections draw upon Section 3.0, where a foundation for analysis of decoupling is developed. The designer can accept the minor risk of using the results presented in Sections 2.0 or 4.0 without reference to the mathematics of Section 3.0, or analyze each case by using the approach outlined in Section 3.0.

In Section 3.0, a numerical example is carried throughout the theoretical development to indicate the order of magnitude of "typical" results for a good design. The reader should not let that example cloud the fact that the analytical methodology is applicable to a wide variety of PCB layouts, values of decoupling capacitors, etc., which are not illustrated in that example. The analysis is a general one, not confined to the particular layout nor logic family used for exemplary purposes.

Appendix 4 provides results for PCB trace inductance. Those data are useful in PCB layout and design, and are more accurate than are usually found.

# 2.0 SUMMARY

## 2.1 Nature of Principal Results

Emissions considerations demand higher-quality decoupling than susceptibility considerations. The most important recommendations in this book pertain to emissions and are believed to be

- Use a ferrite bead at the power entry point. For large PCBs, several small beads or one large bead may be necessary to maintain bead resistance at the direct currents encountered on those PCBs.

- Thoroughly specify power-entry and decoupling capacitors and ferrite beads. In particular, the equivalent series resistance and inductance of capacitors is as important as the values of those capacitors. Quantitative limits for parasitic inductance and resistance are recommended here; they apply to the large majority of designs, and usually will prevent both radiated and conducted emissions problems which could otherwise occur.

- PCB layout is inextricably related to the decoupling problem, as layout determines most of the inductances which affect decoupling. It is not necessary nor, even, desirable to minimize all inductances, as some of those inductances aid in filtering high-frequency noise. Specific layout recommendations are given in Section 4.1.

## 2.2 A Presentation of Principal Results and Recommendations

The mathematics in the next Section 3.0 are formidable to some, and the author suffers no illusions as to the fraction of the engineering population which will endure all details presented therein. Even for readers interested in the details, it is helpful to present a synopsis of the results now to make the purposes of the analysis obvious from the start.

Reproductions of the transparencies used in the "Decoupling" and "PCB Layout"* parts of the TKC Seminar, **Digital Design for Interference Specifications,** are presented in the remaining pages of this Section - without comment. Section 4.0 provides the same results in more conventional form with supporting text.

*Clock trace layout transparencies are omitted here. Representative clock trace layouts are given in [1].

TKC

DECOUPLING AND LAYOUT OF
DIGITAL PRINTED CIRCUITS

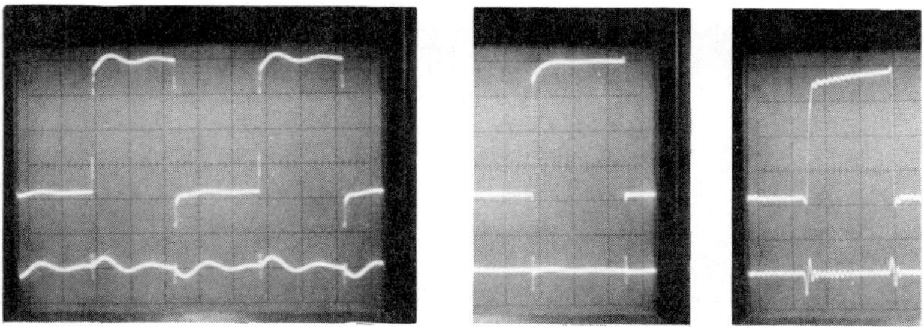

DECOUPLING IS THE SCIENCE AND ART OF DISSOCIATING
CIRCUIT FUNCTIONS FROM THE POWER BUS

TKC • THE KEENAN CORPORATION
Consulting Electronics Engineers

1K2SEM.485

TKC

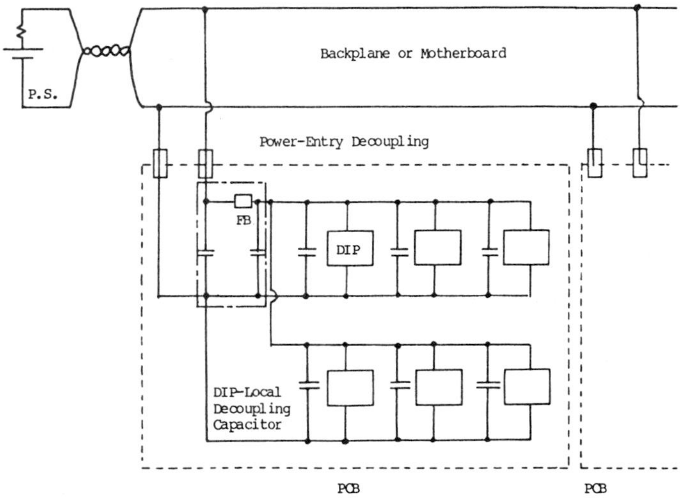


MACROSCOPIC OVERVIEW OF THE DECOUPLING PROBLEM

TKC • THE KEENAN CORPORATION
Consulting Electronics Engineers

SUBJECT4.884

TKC

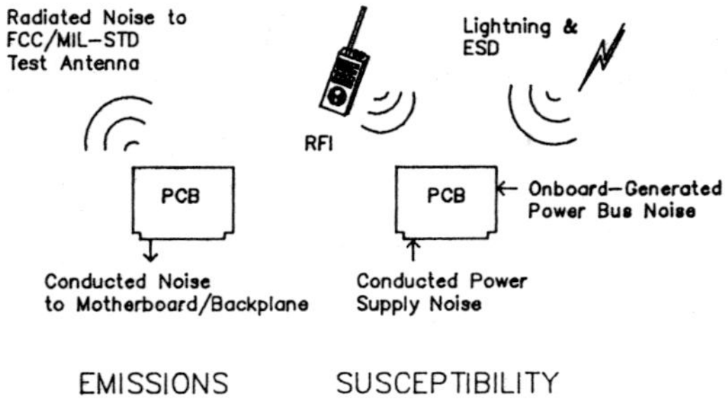


DECOUPLING AFFECTS EMISSIONS AND SUSCEPTIBILITIES

TKC • THE KEENAN CORPORATION
Consulting Electronics Engineers

2K2SEM.485

TKC

- EMISSIONS: PULSE CURRENT IN POWER BUS
- SUSCEPTIBILITY: VOLTAGE EXCURSIONS ON POWER BUS CAN CAUSE LOGIC "FLAKYNESS"
- RELIABILITY: WE ARE ADDING COMPONENTS THAT CAN FAIL
- COST: THOSE COMPONENTS COST MONEY
- TECHNICAL PROBLEMS
  - LAYOUT
  - CAPACITOR QUALITY
  - FERRITE BEAD SELECTION

ASPECTS OF THE DECOUPLING PROBLEM

TKC • THE KEENAN CORPORATION
Consulting Electronics Engineers

ASPECTS.884

TKC

- EMISSIONS SPECIFICATIONS
  - FCC/COMPUTING DEVICES AND SIMILAR EUROPEAN (VDE)
  - MIL-STD-461A/B
  - TEMPEST (NACSIM 5100A)
- SUSCEPTIBILITY SPECIFICATIONS
  - VERY FEW COMMERCIAL SPECS (NO FCC, BUT ...), BUT FIELD RELIABILITY SUBSTANTIALLY AFFECTED
  - MIL-STD-461A/B

SPECIFICATIONS IMPACTED BY DECOUPLING

TKC · THE KEENAN CORPORATION
Consulting Electronics Engineers

3K2SEM.485

TKC

- CAPACITOR MANUFACTURER'S RECOMMENDATIONS
- DIP MANUFACTURER'S RECOMMENDATIONS
- CUT AND TRY
- "PURCHASING, GIVE ME ONE-HUNDRED 0.01 UF'S PER PRODUCT SHIPPED".

HOW MOST HAVE BEEN DECOUPLING

TKC · THE KEENAN CORPORATION
Consulting Electronics Engineers

4K2SEM.485

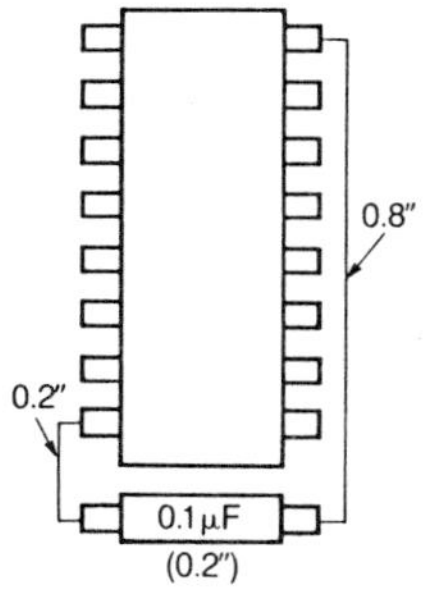


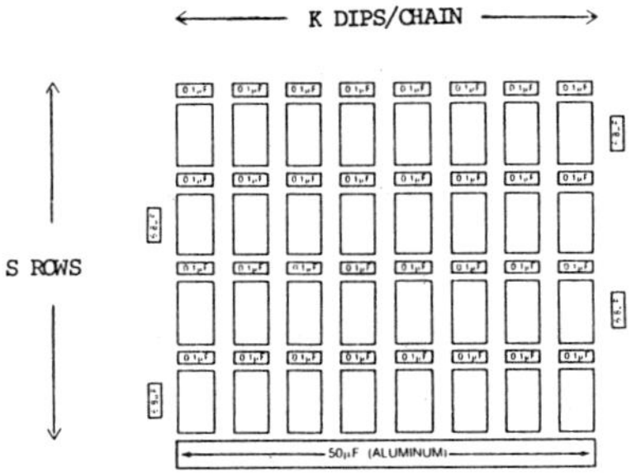


EXCESSIVE DECOUPLING CAP/TRACE INDUCTANCE

TYPICAL BUT UNNECESSARY NUMBER OF ELECTROLYTIC CAPS

TYPICAL, BUT NOT RECOMMENDED BY TKC

TKC • THE KEENAN CORPORATION
Consulting Electronics Engineers

5K2SEM.485

- N = # OF OUTPUT GATES GOING ON $\cong$ ONE-HALF TOTAL # OF OUTPUT GATES
- K = # OF DIPS IN A CHAIN (ROW)
- S = # OF ROWS OF DIPS
- P = # OF DIPS PER DECOUPLING CAPACITOR
- $C_S$ = OUTPUT C OF GATE
- $C_L$ = FANOUT X INPUT C

QUANTITIES USED IN RECOMMENDATIONS

TKC • THE KEENAN CORPORATION
Consulting Electronics Engineers

6K2SEM.485

T
K
C

- CHOOSE TANTALUM OR METALLIZED POLYCARBONATE (NOT ALUMINUM) POWER ENTRY CAPACITORS

  - $C_b \geq 10SKC_d$ • $ESL_b \leq 30$ nH, including leads/traces • $ESR_b \leq 1$ Ohm

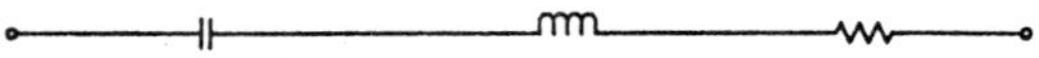

- CHOOSE HIGH-QUALITY DECOUPLING CAPACITORS

  - $C_d \geq 4.5$ x # Output Gates x $(C_s + C_L)$ • $ESL_d < 20$ nH • $ESR_d \leq 0.5$ Ohms

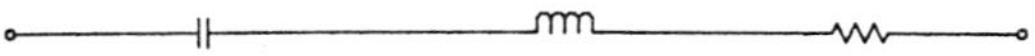

Leads/traces should add no more than 10 nH to $ESL_d$ (alternatively, total $ESL_d$ less than 30 nH). Double $C_d$ if two DIPs are served by one decoupling capacitor. Don't use more than two DIPs per capacitor.

SUMMARY OF RECOMMENDATIONS - PAGE 1

TKC · THE KEENAN CORPORATION
Consulting Electronics Engineers

7K2SEM.485

T
K
C

- USE A FERRITE BEAD!

  - $L_{fb}(f_c)/[R_{fb}(f_c)]^2 < KSC_d/4$, [$f_c < f_d$, $f_c$ = Clock Frequency]
  - 30 Ohms [@ $f > 5f_c$] < $R_{fb}(f)$ < 10 Ohms [@ $f = f_c$]

Note: The above values should be maintained at the direct operating current.

SUMMARY OF RECOMMENDATIONS - PAGE 1A

TKC · THE KEENAN CORPORATION
Consulting Electronics Engineers

7AK2SEM.685

TKC

- LAY OUT THE PCB POWER BUS PROPERLY

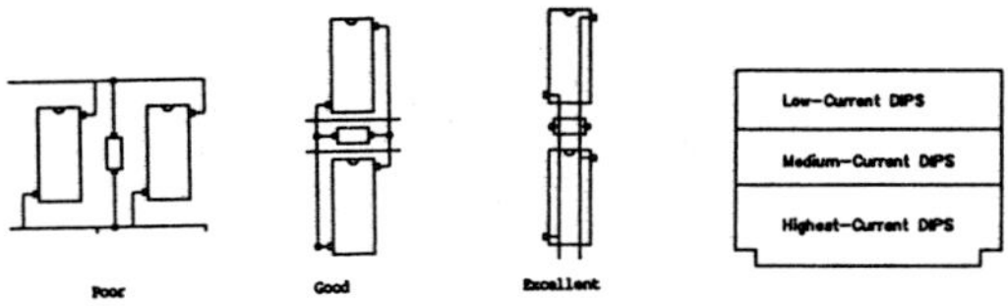


- EXERCISE SPECIAL CARE IN DECOUPLING CLOCK AND I/O DIPS

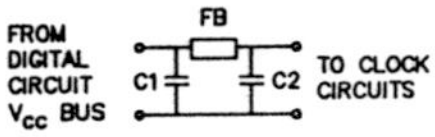


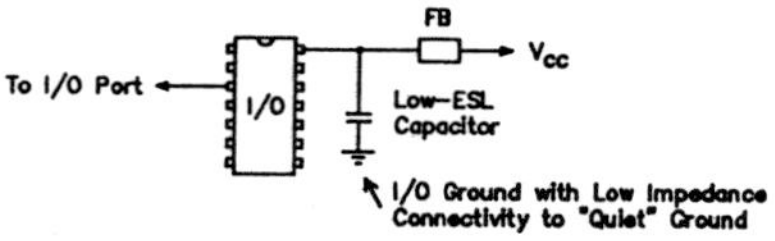


SUMMARY OF RECOMMENDATIONS - PAGE 2

TKC · THE KEENAN CORPORATION
Consulting Electronics Engineers

8K2SEM.485

TKC

- LEAVE AS MUCH COPPER AS POSSIBLE ON THE PCB. CONNECT THOSE COPPER LANDS TOGETHER AND TO THE PCB POWER-ENTRY GROUND, OR

- USE A GROUND GRID, OR

- USE A MULTILAYER PCB WITH A GROUND PLANE (OFTENTIMES IMPRACTICAL)

SUMMARY OF RECOMMENDATIONS - PAGE 3

TKC · THE KEENAN CORPORATION
Consulting Electronics Engineers

9K2SEM.485

T
K
C

- SPECIFYING CAPACITORS BEYOND VALUE AND WORKING VOLTAGE

- USING FERRITE BEADS

- THINKING OF PCB TRACES AS INDUCTORS/TRANSMISSION LINES

- PAYING ATTENTION TO POTENTIAL EMISSIONS PROBLEMS
  EARLY IN THE DESIGN

BREAKS WITH SOME TRADITIONS

TKC • THE KEENAN CORPORATION
Consulting Electronics Engineers

11K2SEM.485

T
K
C

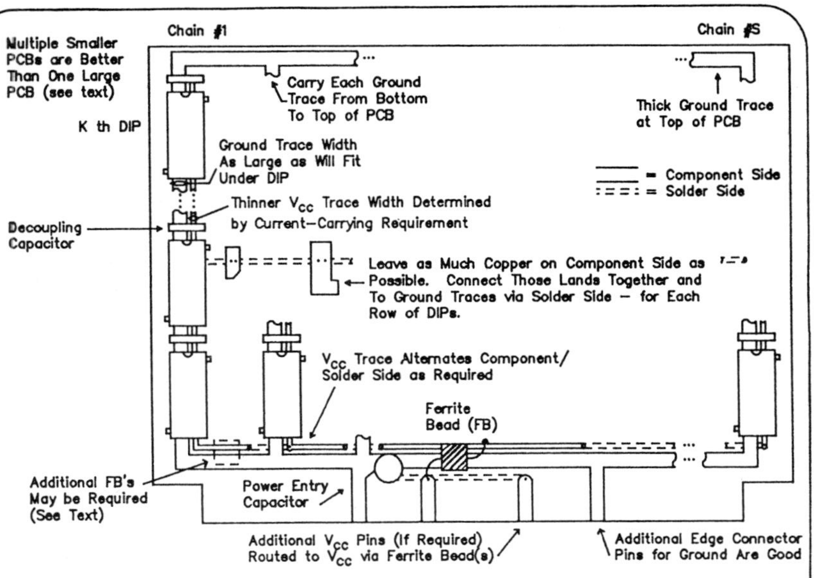


SUMMARY OF PCB POWER BUS LAYOUT RECOMMENDATIONS

TKC • THE KEENAN CORPORATION
Consulting Electronics Engineers

PCBPWR.685

TKC

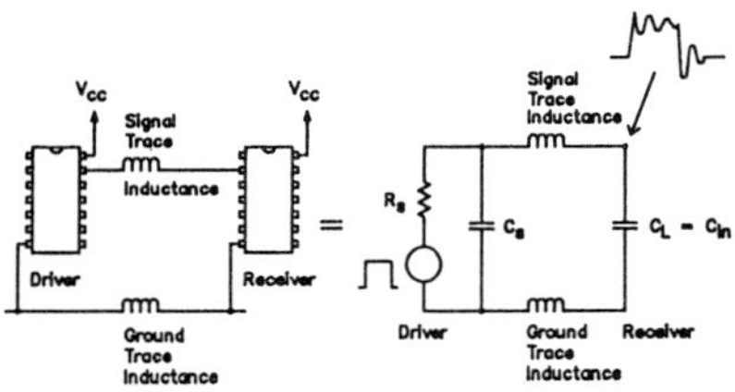


GROUND INDUCTANCE

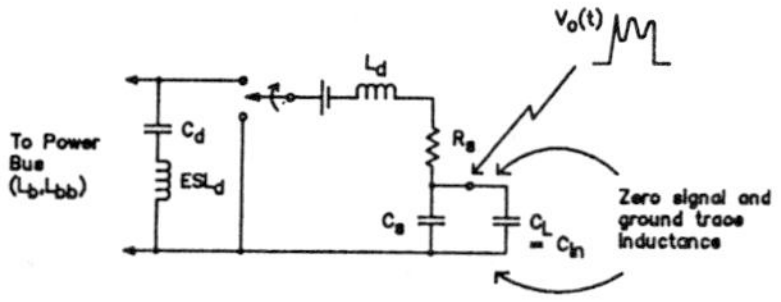


THE DECOUPLING PROBLEM

DIFFERENTIATING BETWEEN GROUND NOISE
AND DECOUPLING PROBLEMS

TKC • THE KEENAN CORPORATION
Consulting Electronics Engineers

12K2SEM.485

TKC

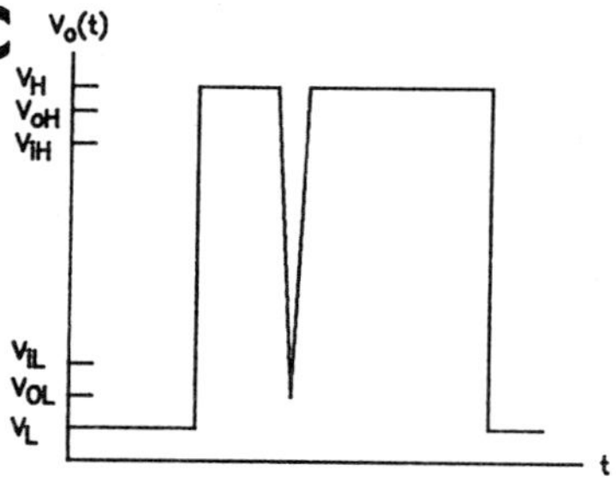


| | $V_{oL}$ | $V_{iL}$ | $V_{oH}$ | $V_{iH}$ | Noise Immunity Low/High |
|---|---|---|---|---|---|
| CMOS [B Series] | 0.05 | 0.3 $V_{CC}$ | $V_{CC}$ -.05 | 0.7$V_{CC}$ | ≅ 0.3$V_{CC}$ |
| HCMOS | 0.10 | 1.0 | 4.9 | 4.5 | 900 mV/400 mV |
| LSTTL | 0.50 | 0.80 | 2.7 | 2.0 | 300 mV/700 mV |
| ECL [10K, 10KH] | -1.63 | -1.48 | -0.98 | -1.1 | 150 mV/120 mV |

$V_{oL}$ = Maximum gate output voltage, low state

$V_{iL}$ = Maximum gate input voltage, low state

$V_{oH}$ = Minimum gate output voltage, high state

$V_{iH}$ = Minimum gate input voltage, high state

AN OUTPUT GLITCH

NOISE IMMUNITY CRITERIA

SUSCEPTIBILITY

TKC • THE KEENAN CORPORATION
Consulting Electronics Engineers

13K2SEM.485

TKC

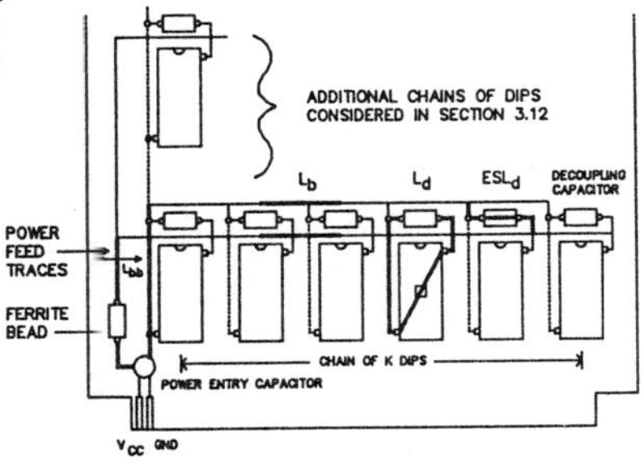


Traces: ——— = Component Side - - - - = Solder Side. FB = Ferrite Bead.

K = # OF DIPS (= 6)

$L_{BB}$ = INDUCTANCE OF POWER FEED TRACES (= 23.1 NH)

$L_B$ = INDUCTANCE OF DIP-TO-DIP POWER BUS SEGMENT (= 13.3 NH)

$ESL_D$ = CAP & LEADS & TRACES TO POWER BUS, NOT DIP (= 24.8 NH)

$L_D$ = GATE-TO-DECOUPLING CAP (= 53.5 NH)

A CONVENIENT EXAMPLE



TKC

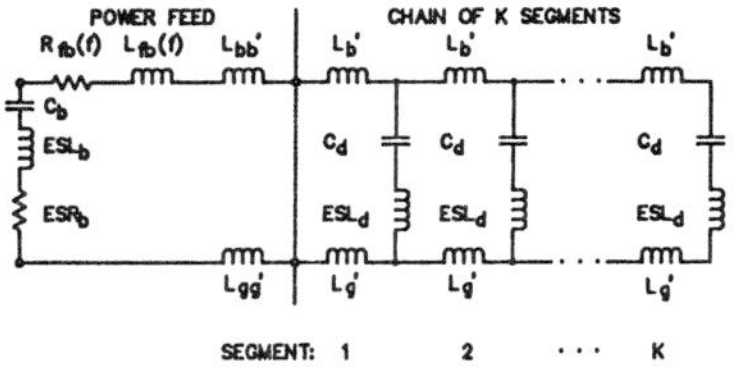


REAL WORLD

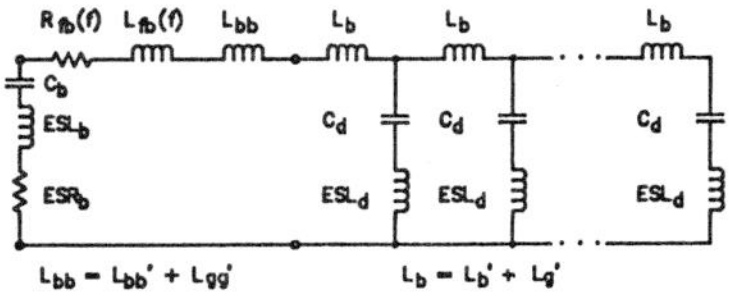


CONVENIENT ANALYSIS - ALL SEGMENT VOLTAGES, ETC., REFERENCED TO DECOUPLING CAPACITOR GROUND

EQUIVALENT CIRCUITS OF THE POWER BUS

TKC

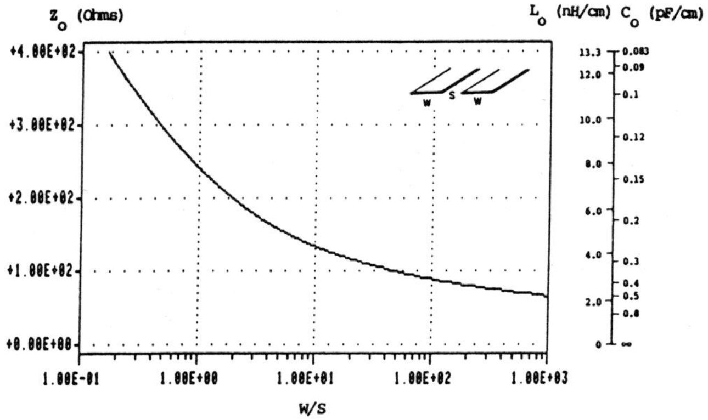


HOW TO DETERMINE INDUCTANCES



TKC

- AT HIGH FREQUENCIES, CASCADE OF

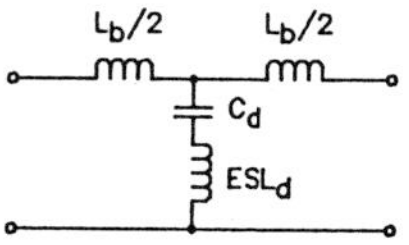


BEHAVES LIKE

- AT LOW FREQUENCIES, SAME CASCADE BEHAVES LIKE

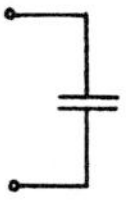

$$\frac{KC_d}{[1 - (f/f_d)^2]^{1/2}}$$

DISTILLED ESSENCE OF SOME TRICKY MATH

- HIGH FREQUENCIES

$$F > \frac{F_D}{[1 + L_B/4ESL_D]^{1/2}}, \quad F_D = \text{SELF-RESONANCE OF DECOUPLING CAP}$$

AND POWER BUS SEGMENTS ARE ISOLATED FROM EACH OTHER. HF NOISE ATTENUATED AS IT MOVES TOWARDS POWER ENTRY POINT.

- LOW FREQUENCIES

$$F < \frac{F_D}{[1 + L_B/4ESL_D]^{1/2}}$$

AND POWER BUS SEGMENTS LUMP TOGETHER. NO NOISE ATTENUATION IN K-DIP CHAIN. SUPPRESSION VIA FERRITE BEAD AND POWER ENTRY CAPACITOR.

## HIGH AND LOW FREQUENCIES

TKC • THE KEENAN CORPORATION
Consulting Electronics Engineers

19K2SEM.485

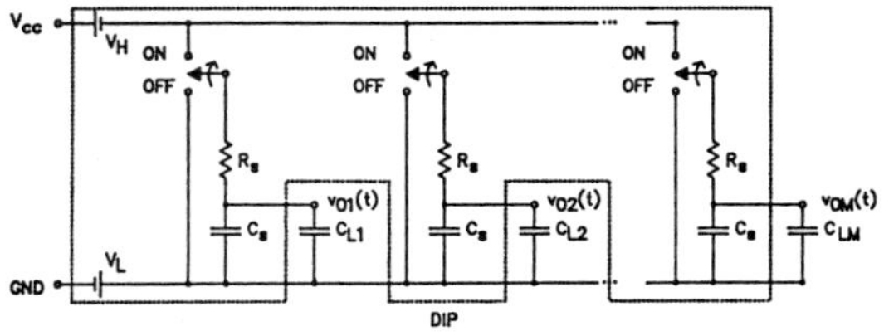


A. Single M-Output Gate DIP and Loads $C_{L1}$, $C_{L2}$, ... $C_{LM}$

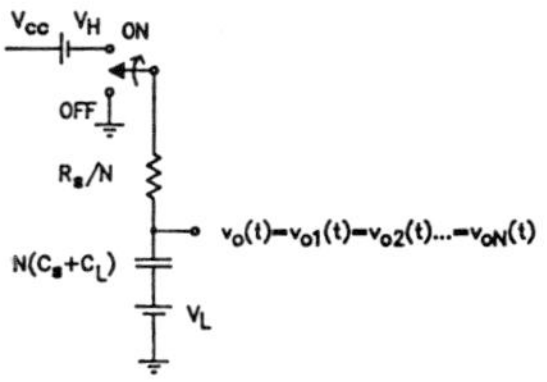


B. DIP Equivalent Circuit for N Equally-Loaded Output Gates Going On

## EQUIVALENT CIRCUIT OF A DIP

TKC • THE KEENAN CORPORATION
Consulting Electronics Engineers

17K2SEM.485

T
K
C

| | CMOS [B Series] | | HCMOS*** | LSTTL | ECL [10K, 10KH] |
|---|---|---|---|---|---|
| $\Delta$ V Out* | 5.0 | 15.0 | 5.0 | 3.3 | 1.0 |
| $\tau_r$**, ns | 80 @ 50 pF | 20 @ 50 pF | 8 @ 50 pF | 17.6 @ 15 pF | 3.5 @ 50 Ohms |
| $\tau_f$**, ns | 80 @ 50 pF | 20 @ 50 pF | 8 @ 50 pF | 4.5 @ 15 pF | 3.5 @ 50 Ohms |
| $R_s$, Ohms | 600 | 160 | 45 | 228/51 N/F | 7 |
| $C_s$, pF | 10 | 6 | 30 | 20/25 N/F | 13 |
| $R_{in}$, Ohms | > 1 M | | > 1 M | 100 K | 1.3 K |
| $C_{in}$, pF | 5 | 5 | 4 | 5 | 3 |

* "Typical" output swing. Guaranteed swings are implicit in Table 3.4-1.
** Transition $\tau$'s are for 10 % to 90 % points. @ indicates external loading.
*** Due to possible overlapped "on" conditions in the internal totem-pole output circuit, the model is only approximately applicable to HCMOS and TTL.

REPRESENTATIVE LOGIC FAMILIES

TKC • THE KEENAN CORPORATION
Consulting Electronics Engineers

18K2SEM.485

T
K
C

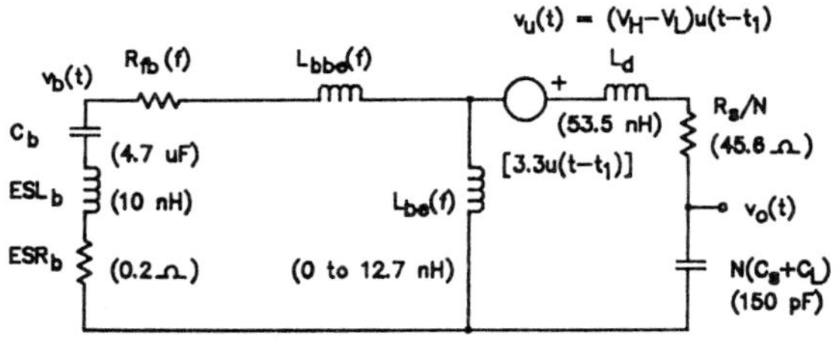


HIGH-FREQUENCY MODEL (DISCHARGING)

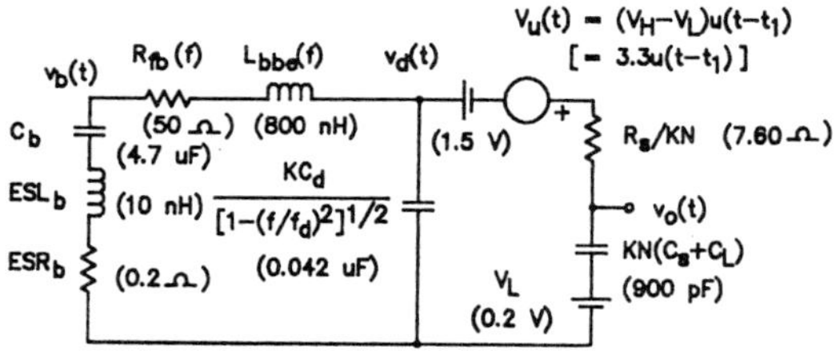


LOW-FREQUENCY MODEL (CHARGING)

HIGH AND LOW FREQUENCY MODELS

TKC • THE KEENAN CORPORATION
Consulting Electronics Engineers

20K2SEM.485

TKC

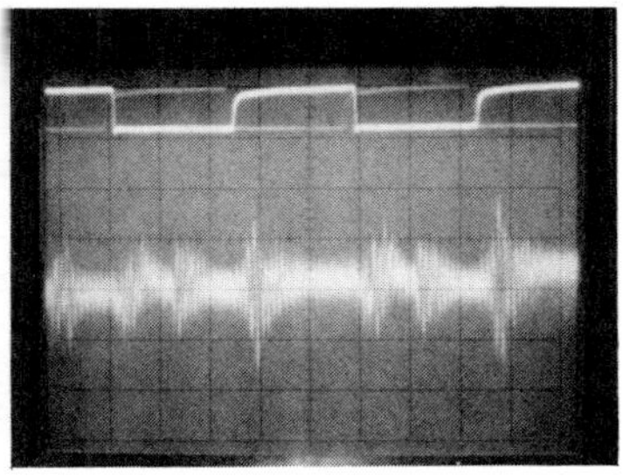

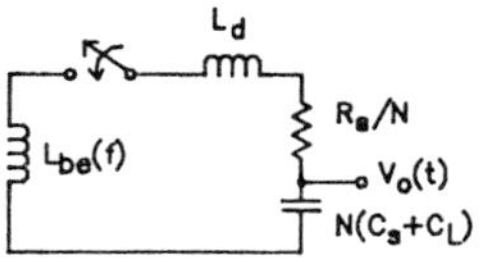


POPULAR HOME COMPUTER (200 NS/CM, 10 MV/CM)

**DECOUPLING CAP NOISE** **MODEL**

END SEGMENT IS WORST CASE FOR HF RINGING

TKC · THE KEENAN CORPORATION
Consulting Electronics Engineers

21K2SEM.485

TKC

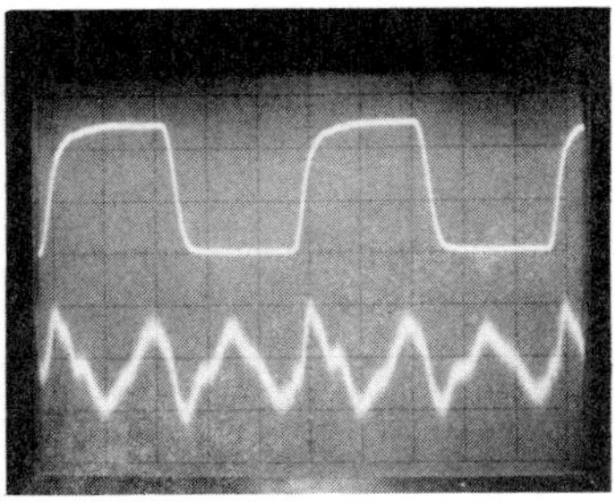

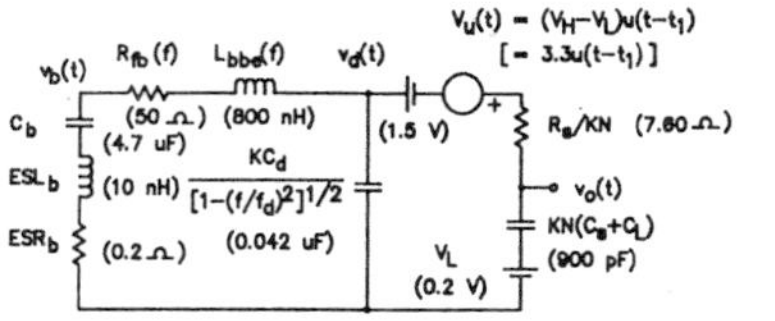


Values shown are for LSTTL, K = 6, N = 5, $C_L$ = 10 pF (Fanout = 2/Gate).

Z80A IN POS TERMINAL (50 NS/CM, 50 MV/CM)

(NOTE: 20 MHZ BW LIMIT)

**DECOUPLING CAP NOISE** **MODEL**

ALL DIPS IN CHAIN PARTICIPATE IN LF RINGING

TKC · THE KEENAN CORPORATION
Consulting Electronics Engineers

22K2SEM.485

TKC

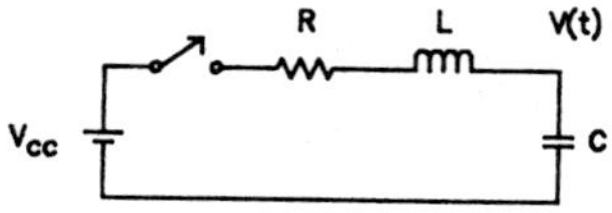


If $D = \dfrac{4L}{R^2C} > 1$, the circuit is <u>UNDERDAMPED</u> and rings:

$$v(t) = V_{CC}[1 - e^{-at}(\cos Bt + (a/B)\sin Bt)],\ \text{where}$$

$$a = R/2L \qquad \text{and} \qquad B = |1/LC - a^2|^{1/2}$$

If $D = \dfrac{4L}{R^2C} < 1$, the circuit is <u>OVERDAMPED</u> and does not ring:

$$v(t) = V_{CC}[1 - e^{-at}(\cosh Bt + (a/B)\sinh Bt)],\text{where}$$

a and B are as given above. If $D \ll 1$ (strong overdamping),

$$v(t) \cong V_{CC}[1 - e^{-t/RC}],\ t \gg 1/RC$$

FUNDAMENTALS OF DAMPING/RINGING

TKC · THE KEENAN CORPORATION
Consulting Electronics Engineers

23K2SEM.485

TKC

<---------------------------- NOISE

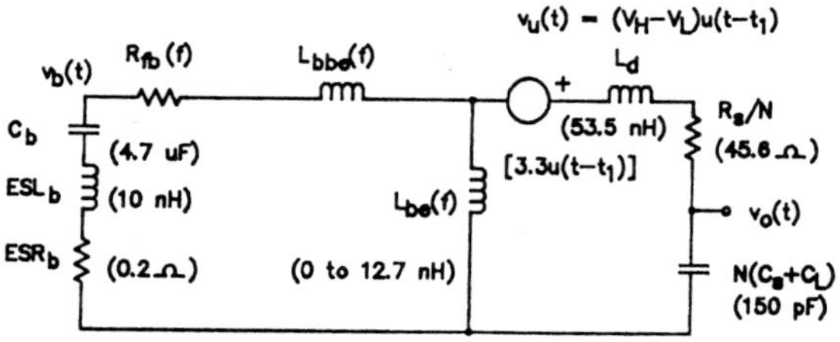


TRANSFER OF HIGH-FREQUENCY SWITCHING NOISE TO POWER-ENTRY POINT

TKC · THE KEENAN CORPORATION
Consulting Electronics Engineers

6K2SEM.685

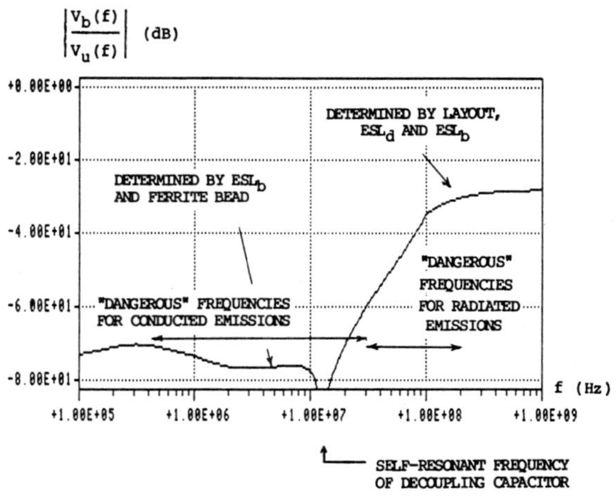


CONDUCTED NOISE FOR A WELL-DECOUPLED CHAIN OF DIPS

TKC · THE KEENAN CORPORATION
Consulting Electronics Engineers

7K2SEM.685

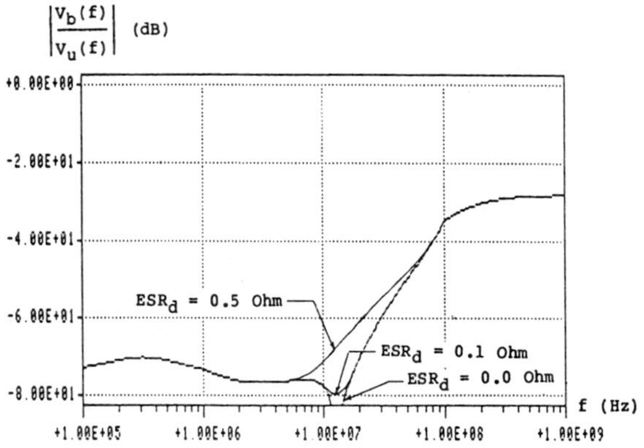


EFFECT OF NONZERO DECOUPLING CAPACITOR
EQUIVALENT SERIES RESISTANCE ($ESR_D$)

TKC · THE KEENAN CORPORATION
Consulting Electronics Engineers

3K2SEM.685

T
K
C

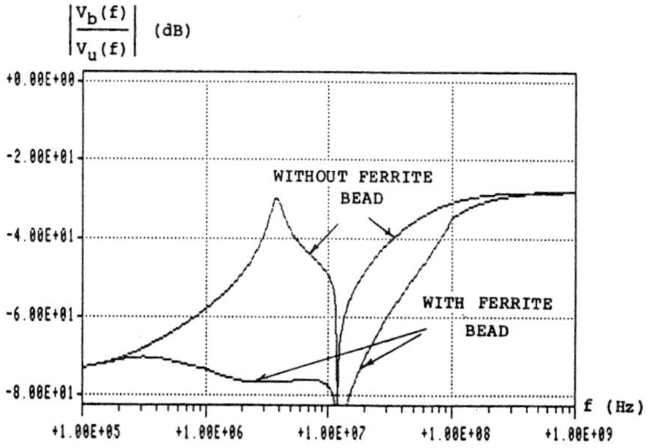


CONDUCTED NOISE WITH AND WITHOUT A FERRITE BEAD

TKC · THE KEENAN CORPORATION
Consulting Electronics Engineers

2K2SEM.685

T
K
C

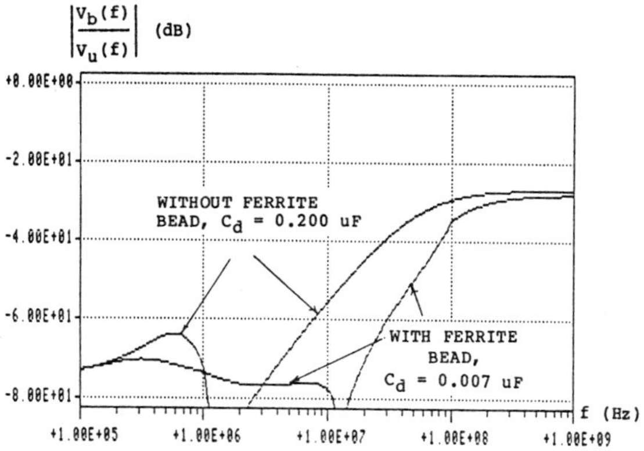


ONE ATTEMPT TO GET ALONG WITHOUT A BEAD

TKC · THE KEENAN CORPORATION
Consulting Electronics Engineers

1K2SEM.685

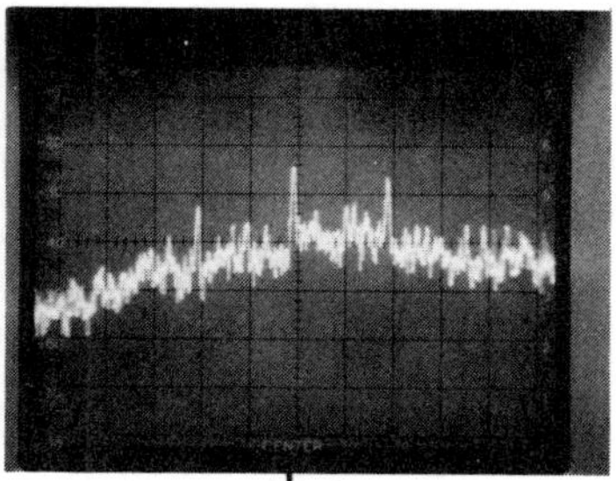

60 MHz

10 DB/CM, 2 MHZ/ CM
RADIATED NOISE

- DIRECT LOOP RADIATION FROM PCB (USUALLY SMALLER THAN FROM CLOCK LOOPS)
- BACKPLANE RADIATION
- CONDUCTED NOISE TO I/O DIPS TO OUTPUT PORTS*
- TO THE POWER CORD*

* THE REALLY BAD GUYS

RINGING CAUSES RADIATED EMISSIONS

TKC · THE KEENAN CORPORATION
Consulting Electronics Engineers

25K2SEM.685

---

- LOOP ANTENNAS

$$E = \text{CONSTANT} \times \frac{I(F)\ F^2(\text{AREA})}{\text{MEASUREMENT DISTANCE}}$$

- WHIP ANTENNAS

$$E = \text{CONSTANT}_2 \times \frac{I(F)\ F^2(\text{LENGTH})}{\text{MEASUREMENT DISTANCE}}$$

- AS F IS INCREASED, I(F) DECREASES. USUAL WORST-CASE FREQUENCY RANGE FOR DIGITAL LOGIC IS

  $50\ \text{MHZ} < F < 150\ \text{MHZ}$

  ALSO, WORST SUSCEPTIBILITY PROBLEMS

"ANTENNA" EFFICIENCY

TKC · THE KEENAN CORPORATION
Consulting Electronics Engineers

26K2SEM.485

TKC

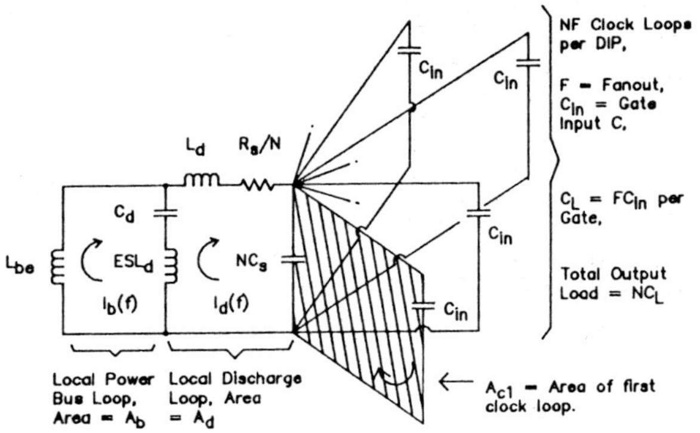


WHY CLOCK TRACE RADIATION IS WORSE THAN THAT FROM THE POWER BUS



TKC

- LET

  N = # ON-GOING GATES

  P = # DIPS PER DECOUPLING CAPACITOR

- THEN

$$NP < \frac{R_S^2 \, (C_S + C_L)}{4 \, (L_{BE}(F) + L_D/P)}$$

- USUALLY RESTRICTS TO MAX P = 2

PREVENTION OF DISCHARGE LOOP RINGING

- MODEL

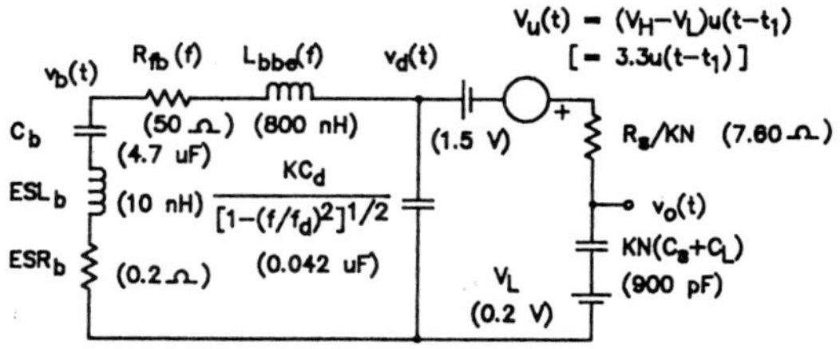


- TO ACHIEVE OVERDAMPING,

$$C_d > \frac{4(L_{bbe}(f) + ESL_b)[1 - (f/f_d)^2]^{1/2}}{K(R_{fb}(f) + ESR_b)^2} = \begin{cases} 200 \text{ pF with ferrite bead,} \\ 0.158 \text{ uF without ferrite bead } \{L_{fb}(f) = R_{fb}(f) = 0\}, \end{cases}$$

PREVENTION OF CHARGING LOOP RINGING

TKC • THE KEENAN CORPORATION
Consulting Electronics Engineers

29K2SEM.485

- LF RINGING EASILY SOLVED WITH FERRITE BEAD

- HF RINGING SOLVED BY
  - PCB LAYOUT (LOW $L_D$, $ESL_D$)
  - NOT TRYING TO OVERLOAD ONE DECOUPLING CAPACITOR WITH TOO MANY DIPS

STOP RINGING AND SOLVE EMISSIONS PROBLEMS

(THAT'S WHAT THE RECOMMENDATIONS DO!)

TKC • THE KEENAN CORPORATION
Consulting Electronics Engineers

30K2SEM.485

TKC

- FERRITE BEADS

- DECOUPLING CAPACITORS

SOME NOTES ON COMPONENT SELECTION FOR DECOUPLING



TKC

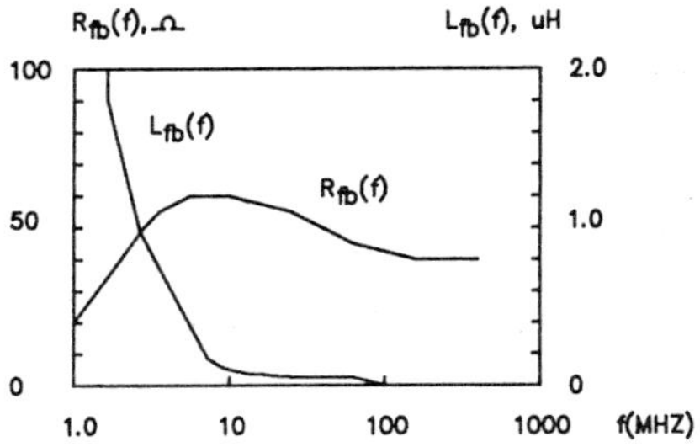


THE IDEAL FERRITE BEAD HAS NO INDUCTANCE
AND SUBSTANTIAL RESISTANCE AT HF

TKC

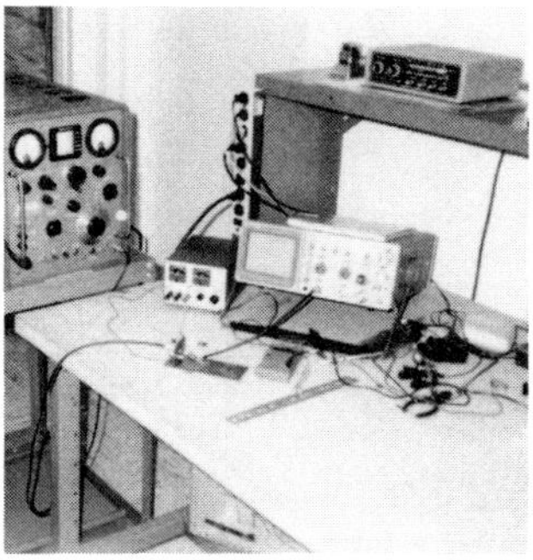

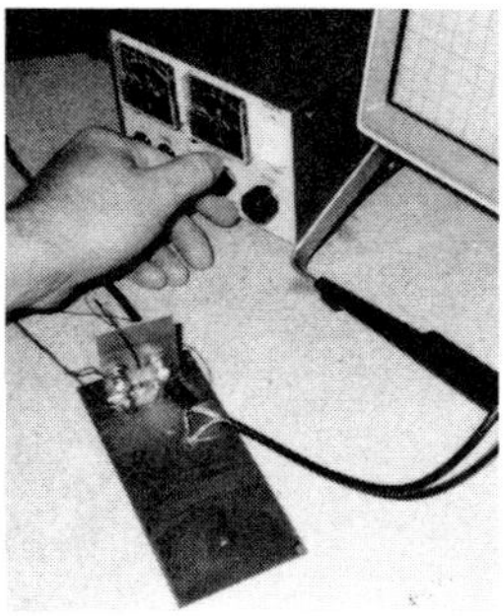

## MEASUREMENT OF EFFECTS OF CORE SATURATION

TKC
FB12.384

TKC

INDIANA GENERAL

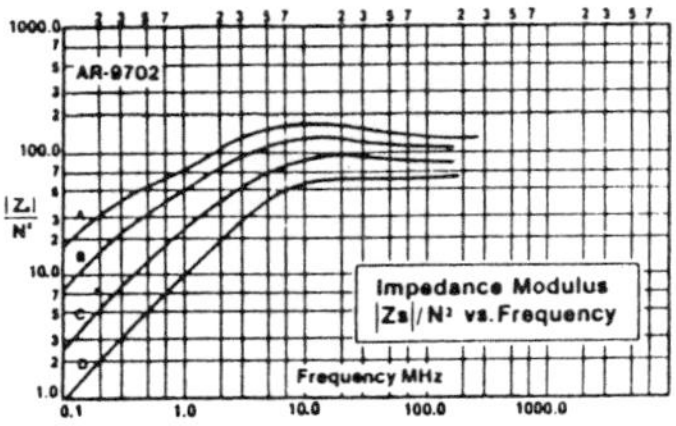


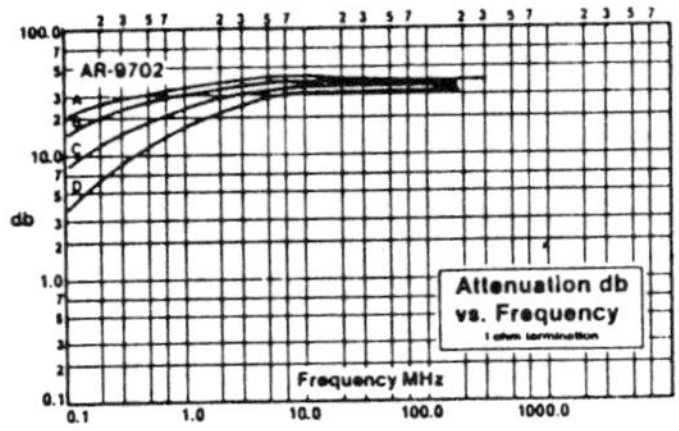


# AR-9702

### Mechanical Specifications

| |
|---|
| $d_1$ = 0.200 ± 0.005 in |
| $d_2$ = 0.050 ± 0.005 in |
| h = 0.400 ± 0.010 in |

INS.

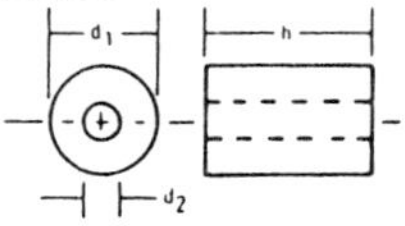

### Electrical Specifications

| D.C. Amp Turns | $\lvert Z_s \rvert / N^2$ 50 kHz | $\lvert Z_s \rvert / N^2$ 100 MHz |
|---|---|---|
| 0.0 Amp Turns | 6.43 Ωmin. | 106 Ωmin. |
| 0.7 Amp Turns | 0.322 Ωmin. | 50 Ωmin. |

Note: Measured with 1 turn < 5 mV

### Ampere Turn Parameters

(A) 0.0 NI (A turns)
(B) 0.17 NI (A turns)
(C) 0.37 NI (A turns)
(D) 0.70 NI (A turns)

## IMPEDANCE DECREASES WITH CURRENT

TKC
FB8.384

TKC

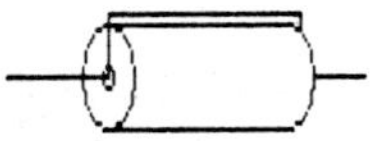

• Two Turns to Boost Impedance

(Note: Multiple turns can lead to prohibitively low self-resonant freq's)

• Cracked (Air-gapped) Bead to Reduce Saturation Effects

(Note: Nobody manufactures them)

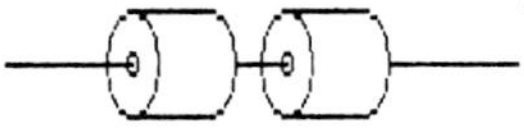

• Multiple Beads to Increase Impedance (Foolish)

## TRICKS OF THE TRADE

TKC
FB5.384

TKC

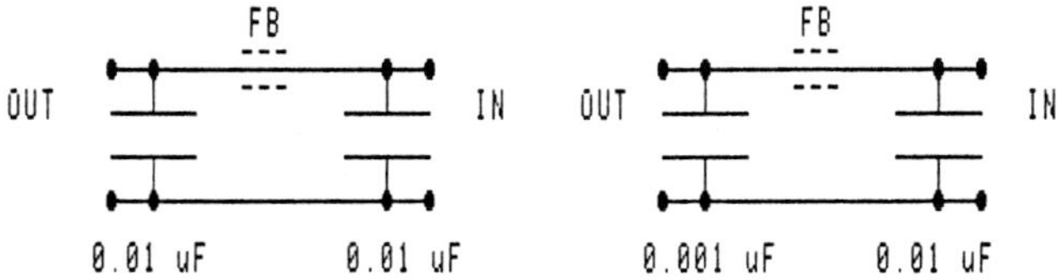


BEAD: Fair-Rite # 2643000301, Two Turns (Idc < 10 mA)
50 mV/cm, 5 usec/cm

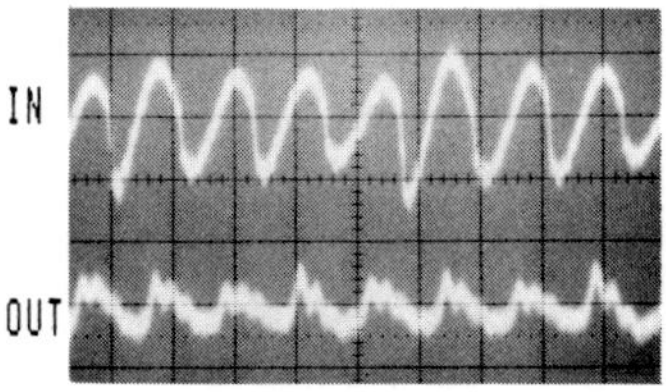


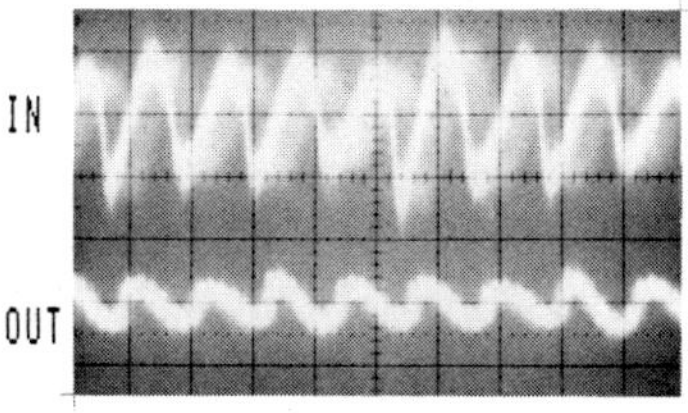


## SUPPRESSION OF NOISE GENERATED BY PCB

TKC
FB4.384

TKC

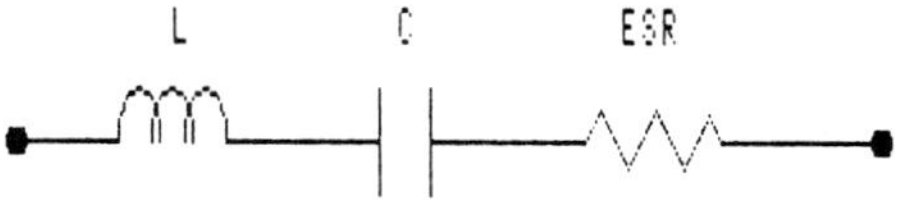


LEAD & PACKAGE INDUCTANCE

CAPACITOR

EQUIVALENT SERIES RESISTANCE (ESR)

$$\text{SELF-RESONANT FREQUENCY} = \frac{1}{2\pi(LC)^{1/2}}$$

# EQUIVALENT CIRCUIT OF A CAPACITOR

TKC
FB6.384

TKC

| CAPACITOR | .01μF | .1μF | .22μF | 1.0μF | 2.2μF |
|---|---|---|---|---|---|
| FILMS: | | | | | |
| BRAND A | 4.3 | 6.0 | — | 11.8 | — |
| BRAND B | 2.8 | 3.0 | 3.0 | 3.5 | — |
| BRAND C | 4.5 | 7.5 | — | 8.8 | — |
| BRAND D | 7.5 | — | 5.8 | — | — |
| TANTALUMS: | | | | | |
| BRAND E | — | 9.5 | — | 9.5 | — |
| ALUMINUMS: | | | | | |
| BRAND F | — | 15.8 | — | 10.0 | — |
| AVX MLC'S | | | | | |
| CONFORMAL RADIALS | — | 1.8 | — | 1.8 | 2.3 |
| MOLDED RADIALS | — | — | — | 1.3 | — |
| DIP'S | — | 1.3 | 1.0 | — | — |
| MOLDED AXIALS | 2.0 | 3.3 | 3.5 | 5.3 | — |

ZERO LEAD LENGTH
INDUCTANCE, NH

| CAPACITOR TYPE | TYPICAL LEAD LENGTH (MM) | LEAD L (NH) | LEAD $V_L$ (MV) | LEAD IMP (Ω) |
|---|---|---|---|---|
| FILMS: | | | | |
| BRAND A | 3.81 | 1.5 | 60 | 0.9 |
| BRAND B | 2.54 | 1.0 | 40 | 0.6 |
| BRAND C | 3.81 | 1.5 | 60 | 0.9 |
| BRAND D | 2.54 | 1.0 | 40 | 0.6 |
| TANTALUMS: | | | | |
| BRAND E | 5.08 | 2.0 | 80 | 1.3 |
| ALUMINUMS: | | | | |
| BRAND F | 4.45 | 1.8 | 70 | 1.1 |
| AVX MLC'S | | | | |
| CONFORMAL RADIALS | 2.54 | 1.0 | 40 | 0.6 |
| MOLDED RADIALS | 2.54 | 1.0 | 40 | 0.6 |
| DIP'S | 5.08 | 2.0 | 80 | 1.3 |
| MOLDED AXIALS | 7.62 | 3.0 | 120 | 1.9 |

TYPICAL LEAD LENGTHS
REQUIRED TO CONNECT

# SOME MORE CAPACITORS

COURTESY AVX CERAMICS

TKC • THE KEENAN CORPORATION
Consulting Electronics Engineers

SUBJECT14.884

TKC

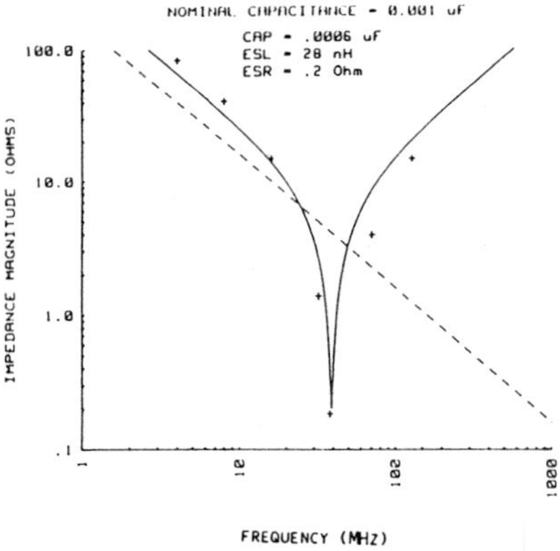


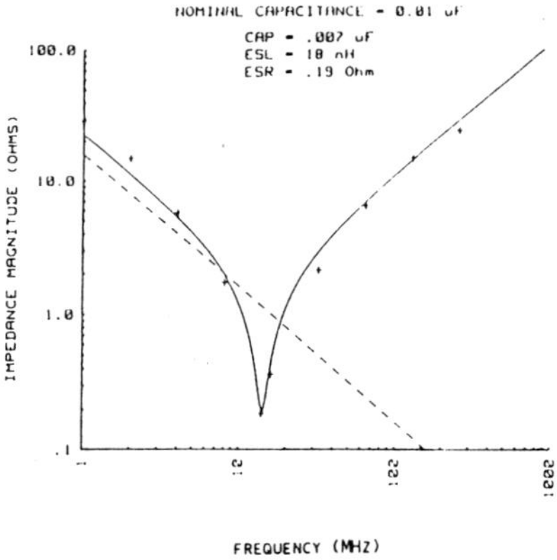


TWO FROM INVENTORY

EACH LEAD = Ø.2"

TKC · THE KEENAN CORPORATION
Consulting Electronics Engineers

35K2SEM.485

TKC

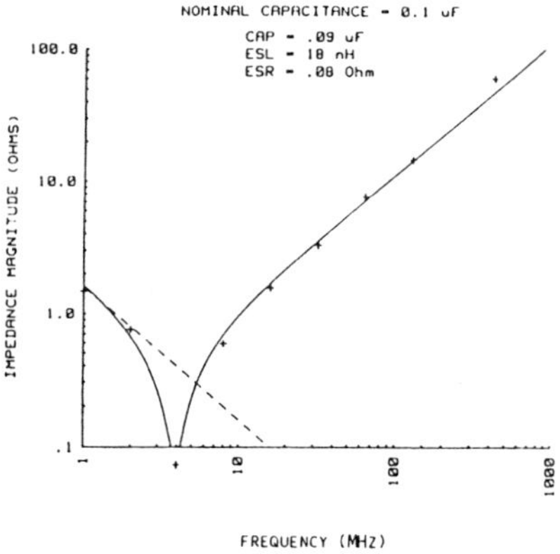


ONE MORE

EACH LEAD = Ø.2"

TKC · THE KEENAN CORPORATION
Consulting Electronics Engineers

34K2SEM.485

T
K
C

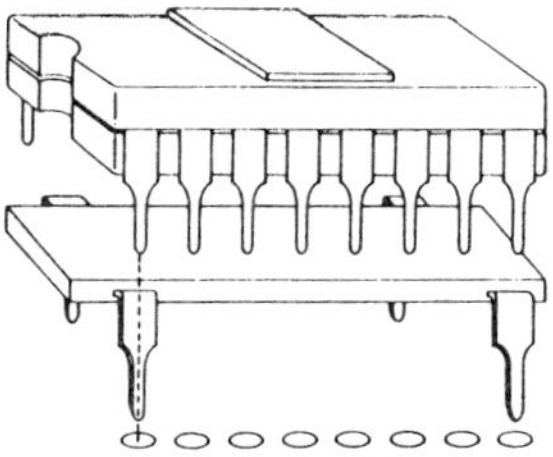

Under-the DIP

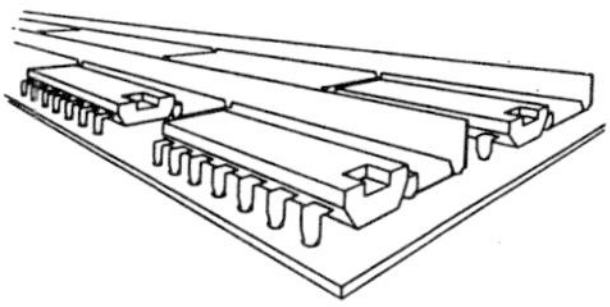

Strip-Type Bus

DRIVES MORE GATES

OK TO INCREASE PCB DENSITY, BUT COULD BE MORE OUTPUT NOISE

SPECIAL POWER BUS PRODUCTS

TKC · THE KEENAN CORPORATION
Consulting Electronics Engineers

33K2SEM.485

# 3.0 ANALYSIS

## 3.1 Symptoms, Description, and Causes of Inadequate Decoupling

The usual manifestations of inadequate decoupling are

- A high level of conducted and/or radiated emissions, and,
- In extreme cases, erratic and/or intermittent circuit operations.

Due to the wide applicability of both military (MIL-STD-461, TEMPEST) and commercial (FCC/VDE) emissions specifications, the former symptom is far more frequently encountered. A system can perform its function well, with no erratic operations, and still substantially exceed an emissions specifications limit.

**Emissions Problems**

There are many case histories of decoupling inadequacies which have caused microcomputer-based products to exceed emissions specifications limits. A few will be cited here:

- A PBX utilized decoupling capacitors which had very large equivalent series inductance. This allowed an excess of noise to be radiated from the physically large power bus backplane, and placed the system's radiation 15 dB over the FCC Class A limit.
- A microcomputer utilized a common digital logic and I/O circuit ground, which placed noise on the cables connected to the I/O ports and caused the system to exceed FCC Class B radiated emissions limits by 10 dB.
- The clock DIP in a data entry product was connected directly to the power bus without benefit of a decoupling resistor/capacitor. Proper decoupling of that clock DIP reduced radiated emissions by 6 dB.

The principles of the above - including the unfortunate and complex connection between noise on the power bus and system radiation - are discussed in Section 3.8.

**Susceptibility Problems**

Erratic operations are far more difficult to cure than emissions problems, as, oftentimes, the occurence of an error depends on gate states; i.e., on what is being processed at a given time. Depending on the frequency of occurrence of errors, it can be difficult to identify and replicate conditions leading to those errors.

The Figure below illustrates noise measured on a 15 VDC power bus of a control system in a nuclear power plant; that noise caused plant shutdowns with a mean-time-between-failure of 17 days. Each shutdown lasted two to four hours, and cost the utility $91,300.00 per hour in lost revenues. In no small measure due to the difficulties of experimentation at the plant, it took eighteen months to finally identify the noise in the Figure as the culprit.

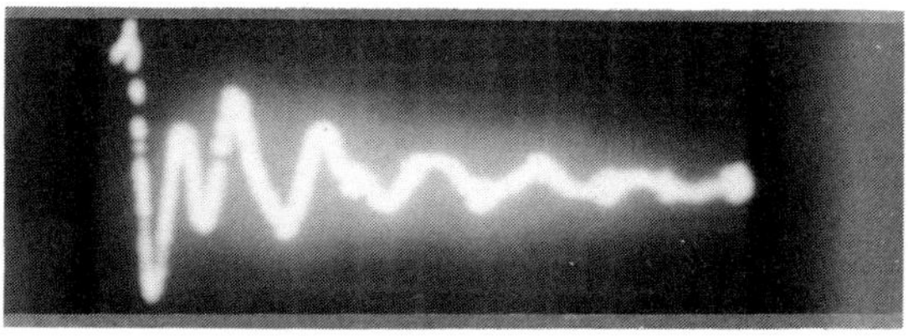

AC coupled scope with Vertical = 1 V/cm, Horizontal = 1 usec/cm

Figure 3.1-1: NOISE ON THE POWER BUS OF A NUCLEAR POWER PLANT CONTROL SYSTEM

Another example of erratic logic operations encountered in practice was that of a kidney dialysis machine which would intermittently malfunction. In contrast to the preceding example, the frequency of failure was tens of minutes, not 17 days, which made diagnosis of the problem much less difficult. After noting excessive noise (1.2 V peak-to-peak) on the 5 VDC bus, the problem was quickly solved with additional decoupling, which reduced the peak-to-peak noise to less than 100 mV.

3.2 Modeling of Power Bus Elements

**Overall Printed Circuit Board (PCB) Power Bus System**

Figure 3.2-1 illustrates a convenient example for the grid-like PCB power bus distribution systems considered here (see Section 4.1 for additional examples).

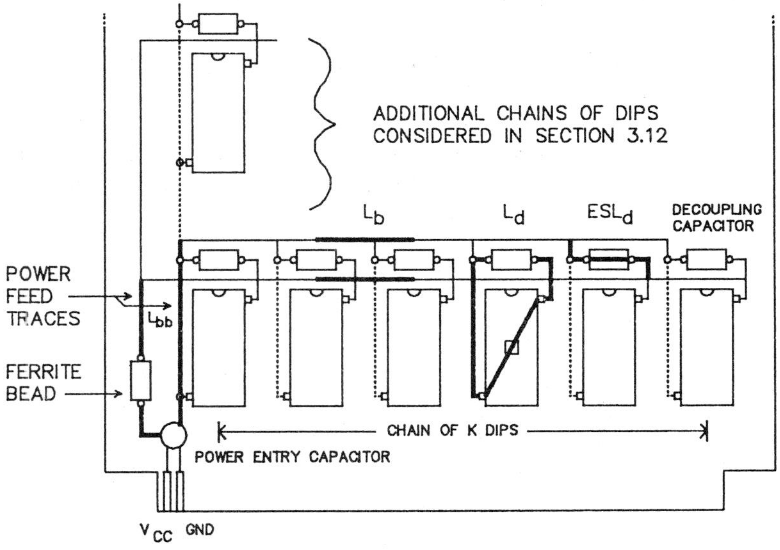


Traces: ------- = Component Side - - - - = Solder Side. FB = Ferrite Bead.

Figure 3.2-1: EXAMPLE OF PCB AND POWER DISTRIBUTION BUS

Initial interest is in the behavior of a single chain of DIPs where

K = Number of DIPs = Number of decoupling capacitors > 1 (= 6), (3.2-1)

$L_{bb}$ = Inductance of power feed traces (= 23.1 nH),

$L_b$ = Inductance of power bus segment (= 13.3 nH),

$ESL_d$ = Internal self-inductance of decoupling capacitor plus that of the leads and traces used to connect to the **power bus** (not the inductance incurred in connecting the capacitor to the DIP) (= 24.8 nH),

$L_d$ = Inductance of traces and intra-DIP wires connecting the gates to the decoupling capacitor (= 53.5 nH).

Numerical data in parentheses are for the layout in the Figure and are derived in Appendix 3. For the present, one DIP per decoupling capacitor is assumed. Possible undesirable consequences of sharing a single decoupling capacitor amongst several DIPs are discussed in Section 3.10. The advantages of using a ferrite bead, and performance without a bead, are discussed in Section 3.6.

**Dual Inline Packages (DIPs) and Loads**

It is assumed that all DIPs in a given chain are of the same type technology (e.g., LSTTL or ECL or CMOS or HCMOS) and that each of the DIPs have one or more output gates which drive input gates, not necessarily within the same logic family. DIPs with additional internal gates are also permitted provided that the usual practical condition prevails: internal gates require much less power than the output gates.

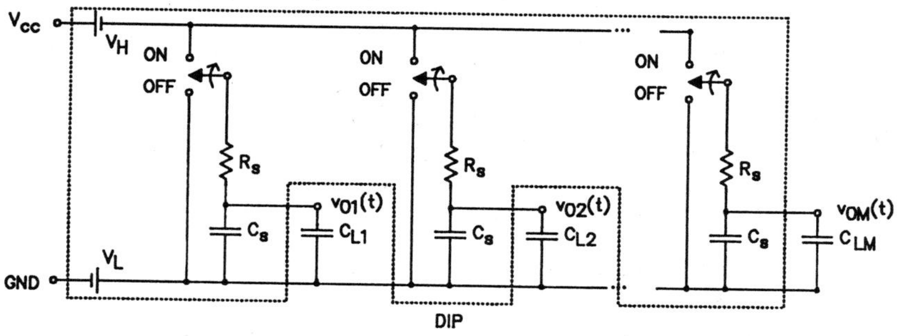


A. Single M-Output Gate DIP and Loads $C_{L1}$, $C_{L2}$, ... $C_{LM}$

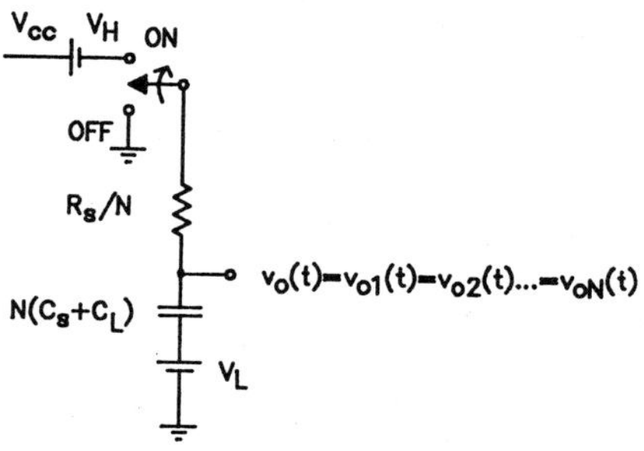


B. DIP Equivalent Circuit for N Equally-Loaded Output Gates Going On

Figure 3.2-2: AC EQUIVALENT CIRCUIT OF A SINGLE DIP

Figure 3.2-2 illustrates the equivalent circuit of a single DIP with M output gates. Figure 3.2-2B is the equivalent circuit of the DIP when N ($\leq$ M) output gates are being switched on and each gate is equally loaded with $C_L = C_{L1} = C_{L2} \cdots = C_{LN}$.

The loads are one or more gates,

$$C_L = \text{Fanout} \times C_{in} = FC_{in}, \qquad (3.2\text{-}2)$$

where $C_{in}$ is the logic-family-dependent input capacitance of the driven gates, exemplified in Table 3.2-1 below. That Table is adapted and expanded from [1, p. 2.2-1]. $\Delta$ V Out = $V_H - V_L$ = Output High minus Output Low Levels.

| | **CMOS [B Series]** | | **HCMOS***** | **LSTTL** | **ECL [10K, 10KH]** |
|---|---|---|---|---|---|
| $\Delta$ V Out* | 5.0 | 15.0 | 5.0 | 3.3 | 1.0 |
| $\tau_r$**, ns | 80 @ 50 pF | 20 @ 50 pF | 8 @ 50 pF | 17.6 @ 15 pF | 3.5 @ 50 Ohms |
| $\tau_f$**, ns | 80 @ 50 pF | 20 @ 50 pF | 8 @ 50 pF | 4.5 @ 15 pF | 3.5 @ 50 Ohms |
| $R_s$, Ohms | 600 | 160 | 45 | 228/51 N/F | 7 |
| $C_s$, pF | 10 | 6 | 30 | 20/25 N/F | 13 |
| $R_{in}$, Ohms | > 1 M | | > 1 M | 100 K | 1.3 K |
| $C_{in}$, pF | 5 | 5 | 4 | 5 | 3 |

* "Typical" output swing. Guaranteed swings are implicit in Table 3.4-1.
** Transition $\tau$'s are for 10 % to 90 % points. @ indicates external loading.
*** Due to possible overlapped "on" conditions in the internal totem-pole output circuit, the model is only approximately applicable to HCMOS and TTL.

Table 3.2-1: CHARACTERISTICS OF REPRESENTATIVE LOGIC FAMILIES

Output source resistance and capacitance, $R_s$ and $C_s$, for representative technologies are shown in Table 3.2-1. For LSTTL, these values change depending on whether the gate is being switched on (N) or off (F). Here, concern is with the "on" values, as it is the pulse of current drawn from the $V_{cc}$ (or $V_{dd}$) bus when the gates are turned on which causes the noise and ringing associated with the $V_{cc}$ (or $V_{dd}$) bus.

No pull-up, pull-down or terminating resistors are included in the equivalent circuits in Figure 3.2-2. The effects of those resistors, if used, are explored in Section 3.9, where it is shown that both ECL and HCMOS logic families rely on those resistors to prevent ringing.

In analyses to follow, it is assumed that none of the other output gates are on at the time of switching, and that all N output gates are simultaneously switched on. Because gates which remain on at the time the other N gates are switched on have a slight filtering/decoupling effect, the condition considered here is a worst case from both emissions and pulse integrity standpoints.

The parasitic inductances associated with the DIP package are omitted in Figure 3.2-2. Although appropriate for the immediately following purposes of developing a model of the power bus, DIP parasitic inductance is not negligible for detailed analysis of either emissions or pulse integrity. DIP parasitic inductance is considered in subsequent more detailed models.

**Power Entry Capacitors**

The power entry capacitor is modeled as

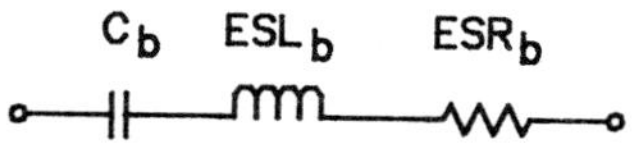


Figure 3.2-3: EQUIVALENT CIRCUIT OF THE POWER ENTRY CAPACITOR

In the Figure,

$C_b$ = Capacitance of power entry capacitor, (3.2-3)

$ESL_b$ = Equivalent series inductance, and

$ESR_b$ = Equivalent series resistance.

In practice, this capacitor is an aluminum or tantalum electrolytic or metallized polycarbonate type. Section 4.2 provides guidelines for selection, wherein it is suggested that aluminum capacitors be excluded from consideration.

**Ferrite Beads**

The equivalent circuit of a ferrite bead is

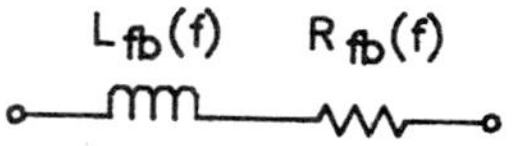


Figure 3.2-4: EQUIVALENT CIRCUIT OF A FERRITE BEAD

In the Figure,

$R_{fb}(f)$ = Frequency-dependent resistance of the ferrite bead, and (3.2-4)

$L_{fb}(f)$ = Frequency-dependent inductance of the ferrite bead.

The bead is selected so that, at VHF (30 - 300 MHz), it is principally resistive with little or no contribution to inductance. The same bead will be inductive at lower frequencies. Acceptable limits of $R_{fb}(f)$ and $L_{fb}(f)$ are given in Section 4.2.

Beads are available in many shapes and sizes with widely varying characteristics [3]. The characteristics of the bead selected for exemplary purposes are given in Figure 3.2-5.

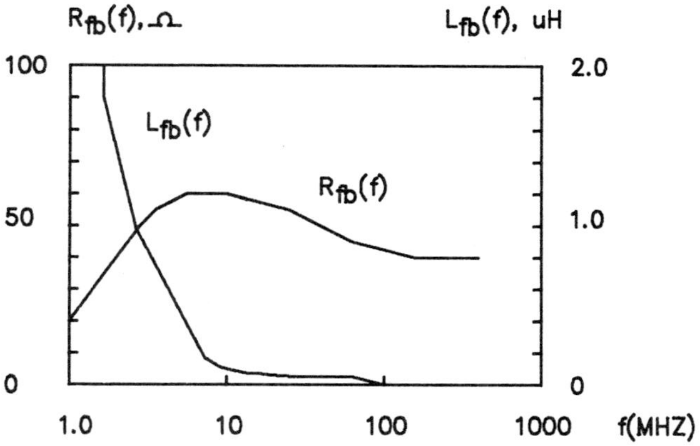


Figure 3.2-5: CHARACTERISTICS OF A FERRITE BEAD AT LOW DIRECT CURRENTS

The reader is referred to [3] for a detailed exposition of the characteristics of ferrite beads, and to Appendix 3 for the basis for derivation of Figure 3.2-5.

**Decoupling Capacitors**

The model for a decoupling capacitor is the same as that given in Figure 3.2-3, with the "b" subscript replaced by "d". In succeeding analyses, equal-value high-Q decoupling capacitors, $C_d$, are assumed:

$$Q = 1/(2\pi f_d C_d ESR_d) >> 1, \quad (3.2\text{-}5)$$

$$f_d = \frac{1}{2\pi [C_d ESL_d]^{1/2}}$$ = Self-resonant frequency of decoupling capacitor,

$ESL_d$ = Internal self-inductance of the capacitor **and that of the leads-/traces connecting the capacitor across the power bus** {not across the DIP; same definition given in (3.2-1)}.

The practical effects of finite-Q (nonzero $ESR_d$) are cited in Section 3.9.

**A Model of the Power Bus**

Initial consideration is given to the power feed and decoupling capacitor segments of the power bus (see, e.g., Figure 3.2-1) without the gates in the DIP on. The general equivalent circuit for any grid-like PCB layout such as that exemplified by Figure 3.2-1 is shown in Figure 3.2-6.

Throughout this work, the worst-case condition of high-impedance power at the PCB entry point is assumed; i.e., it is assumed that the power supply impedance, including that of the backplane or motherboard, is much greater than that of the power entry capacitor and its parasitics. Therefore, no connections other than the power-entry capacitor are shown at the left-hand extreme of the Figure.

In Figure 3.2-6A, the equivalent circuit is shown with the inductance of the power bus distributed between the positive ($V_{cc}$) and ground traces ($L_b'$ and $L_g'$, respectively, for the beginning of the chain of DIPs). It is assumed for now that these inductances are the same between each DIP - i.e., that each DIP is connected to the power bus at equal intervals. However, $L_g'$ and $L_b'$ (and $L_{bb}'$ and $L_{gg}'$) are not required to be equal. Indeed, good design to minimize ground noise (defined in Section 3.4) suggests that $L_g' < L_b'$. Equation (3.2-6) to be presented below allows for two similar traces ($L_g' \cong L_b'$) or a trace plus a ground plane or grid ($L_g' << L_b'$).

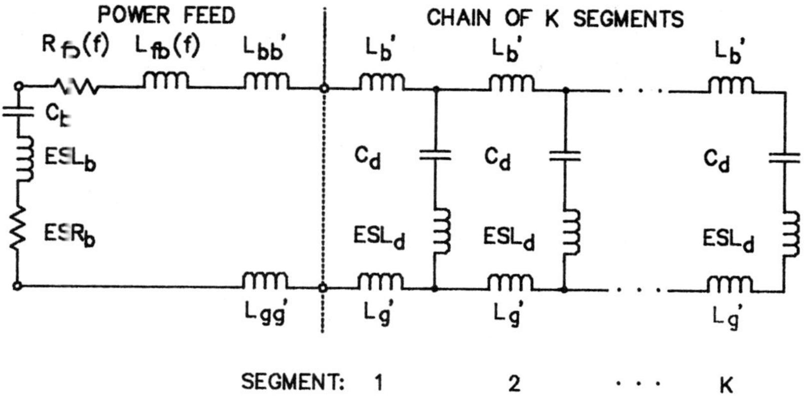


A. Ground and $V_{CC}$ Bus Inductances

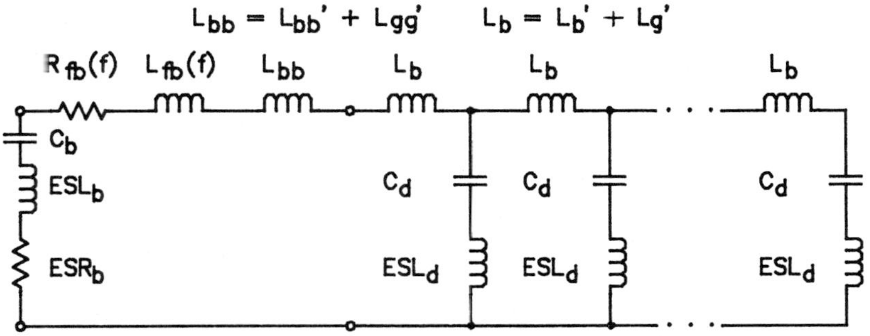


B. Ground and $V_{CC}$ Bus Inductances Lumped Into $V_{CC}$ Side of Bus

Figure 3.2-6: EQUIVALENT CIRCUITS OF THE POWER BUS

In Figure 3.2-6B, inductances are lumped in the $V_{CC}$ leg alone for convenient analysis. This convenience is permitted provided that it is understood that measurements are referenced to the ground end of the decoupling capacitor.

It is clear from Figure 3.2-6A that, due to the drop in voltage in the $L_g$' and $L_{gg}$' parts of the distribution bus, measurements referenced to the ground at the power entry point will not agree with those referenced to the ground end of the decoupling capacitor.

Figure 3.2-6B provides correct results for the voltage and current characterizing a power bus segment provided that measurements are taken with respect to the ground end of the decoupling capacitor in that segment.

**Determination of Inductances**

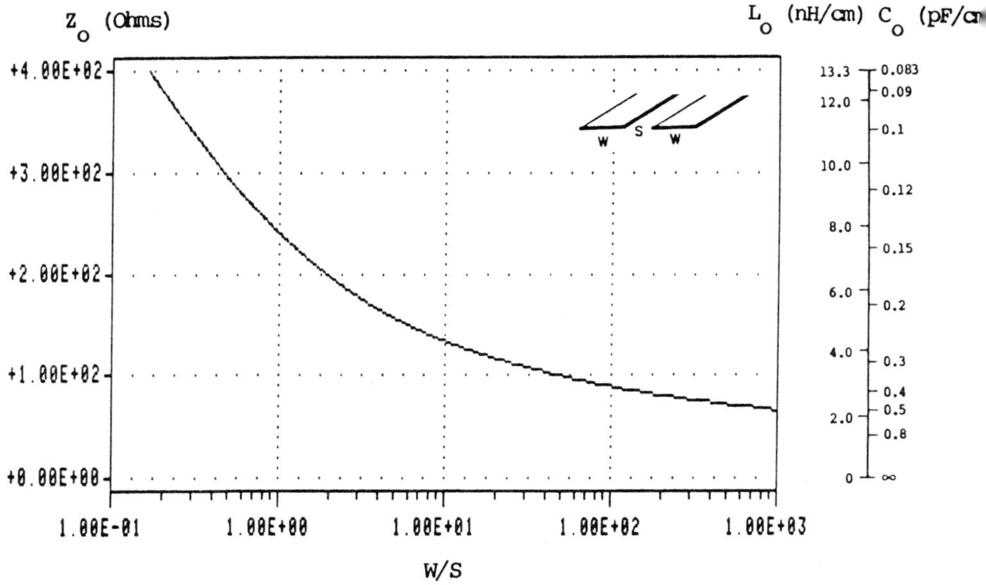


Figure 3.2-7: EXAMPLE OF DETERMINATION OF INDUCTANCES

The inductances associated with the power distribution bus in Figure 3.2-1 can be determined from the exact relationship [2, p. 150]:*

$$L = 0.03333\ Z_o \ \text{nH/cm, where} \qquad (3.2\text{-}6)$$

$Z_o$ = Characteristic impedance (in Ohms) of the power bus traces (transmission lines) determined assuming an air dielectric ($e_r = 1$).

---

* The capacitances associated with the bus traces are omitted in Figure 3.2-6 because, in practical cases, these are much smaller than the lumped capacitors, $C_d$, $C_b$ (Transmission line $C = [33.33(e_r')^{1/2}]/Z_o$ pF/cm [Ibid]).

Typical $Z_o$'s are in the range 100 to 400 Ohms. Then, L = 3.33 to 13.3 nH/cm, respectively, with smaller values corresponding to the use of closely-spaced conductors, such as occurs with ground planes or grids (see Appendix 4). These calculations are exemplified in Appendix 3 for the layout of Figure 3.2-1: the characteristic impedance of the $L_b$ and $L_{bb}$ segments is 315 Ohms, corresponding to 10.5 nH/cm of inductance.

Figure 3.2-7 exemplifies data available for various trace configurations. Appendix 4 provides additional data for other trace/ground plane structures.

Validity of the assumption that PCB traces/planes are purely inductive with inductance given by (3.2-6) is restricted to trace lengths which are much smaller than one-quarter wavelength at the highest frequency of interest (about 300 MHz)*. At 300 MHz, a quarter wavelength in air is 0.25 meters. It is smaller on a PCB:

$$\text{Trace lengths} \ll 0.25/(e_r')^{1/2} \text{ meters} = 9.85/(e_r')^{1/2} \text{ inches,} \quad (3.2\text{-}7)$$
$$\cong 11.2 \text{ cm } (4.4") \text{ to } 20.3 \text{ cm } (8") \text{ in practice,}$$

where $e_r'$ is the **effective** relative dielectric constant. $e_r'$ is between unity and $e_r$, the relative dielectric constant of the PCB material.** $e_r' = e_r$ only when the PCB traces are completely imbedded in the PCB material.

For most PCB configurations, (3.2-7) is not overly restrictive, as most frequencies of interest are below 300 MHz and the effect of a modest violation of (3.2-7) is to reduce the amount of apparent inductance. When the trace segments <u>are</u> one-quarter wavelength long, capacitive loads at the end of the traces appear as inductive loads at the beginning of the traces. The reader is referred to [1, Section 3.0] for a treatment of input impedance of PCB traces as a function of length, loading, and frequency.

---

* For reasons outlined in more detail in [1, p. 3.9-2], radiated emissions from digital circuitry above 300 MHz are seldom a serious problem: the magnitudes of the Fourier components of typical finite transition time clock signals are falling off as $1/f^2$, thereby "evening out" the $f^2$ increase in antenna efficiency cited in (3.8-1). Further, the physical size of typical PCB structures is such that antenna efficiency begins to behave as $f^v$, $v < 2$, at approximately 200 to 300 MHz. In testing/retrofitting over one hundred products, only one out-of-(FCC)-specification occurrence above 300 MHz has been noted.

** In practice, $e_r'$ is, for both coplanar (same side of PCB) and biplanar (opposite sides of PCB) traces, in the range 1.5 to 5. That range has been used in determining the numerical results in (3.2-7).

## 3.3 Models of Power Bus Segments

**Overview**

Mathematical analysis of the K-segment model shown in Figure 3.2-6B is carried out in Appendix 1 using ABCD matrix/Floquet-theoretic techniques. The initial purpose here is to present key qualitative features of the results of that analysis. Detailed referrals to the analysis in Appendix 1 will be given in succeeding subsections.

At high frequencies defined by

$$f > \frac{f_d}{[1 + L_b/4ESL_d]^{1/2}}, \qquad \textbf{(High Frequencies)} \tag{3.3-1}$$

each segment of the power bus is isolated from adjacent segments, and it is analytically appropriate to base predictions of high-frequency performance on consideration of a single power bus segment containing a single decoupling capacitor. At low frequencies defined by

$$f < \frac{f_d}{[1 + L_b/4ESL_d]^{1/2}}, \qquad \textbf{(Low Frequencies)} \tag{3.3-2}$$

the input impedance of the model in Figure 3.2-6B behaves as though the K segments are collapsed into a single capacitor of value K x $C_d$. This reflects a breakdown of the segment-to-segment isolation which is provided by decoupling capacitors and trace inductances at high frequencies.

**Practical Aspects of Low and High Frequency Performance**

High frequencies extend to below the self-resonant frequency of the decoupling capacitor, $f_d$, defined in (3.2-5). In typical good designs, high frequencies begin at 3 MHz to 20 MHz, with significant energy only up to a maximum frequency of approximately 300 MHz. For antenna-efficiency reasons explained in Section 3.8, high frequencies are more troublesome in producing out-of-specification radiated emissions.

Low frequencies - typically below 3 MHz to 20 MHz - are more troublesome from a susceptibility standpoint, as it is those low frequencies which cause the longer-enduring excursions of power bus noise. If of sufficient amplitude, those excursions can cause erratic circuit operations.

**High-Frequency Model** (From Appendix 1)

For high frequencies, the condition

$$e^{K|a|} >> e^{-K|a|}, \qquad (3.3\text{-}3)$$

where

$$a = \text{arccosh}\left[1 - \frac{(2\pi f)^2 L_b C_d/2}{[1 - (f/f_d)^2]}\right], \quad \text{real},$$

prevails in practical circuits. Then, the individual segments are nearly isolated from one another. The first and last segments in the K-segment chain can be modeled as shown in Figure 3.3-1A,C, respectively, with

$$L(f) = \frac{L_b}{2}\left[1 + \left(1 - \frac{4\{1 - [f/f_d]^2\}}{\{2\pi f\}^2 L_b C_d}\right)^{1/2}\right], \qquad (3.3\text{-}4)$$

$$\cong \frac{L_b}{2}\left[1 + \left(1 + \frac{4ESL_d}{L_b}\right)^{1/2}\right], \quad (f/f_d)^2 >> 1.$$

Particularly at frequencies near $f_d$, it can be seen from (3.3-3) that $|a|$ is very large. Therefore, the condition $e^{K|a|} >> e^{-K|a|}$ prevails even for K = 2. A large chain of DIPs need not be assumed, and the beginning and end-segment models shown in Figure 3.3-1 usually apply for a "chain" of 2 DIPs.

Further, for the Pth DIP in from either end of the chain, the model shown in Figure 3.3-1B applies if $e^{P|a|} >> e^{-P|a|}$, as is usually true in practice.

Maximum decoupling is achieved near the self-resonant frequency of the decoupling capacitor. As f approaches infinity, the exponent "a" in (3.3-3) monotonically decreases to arccosh $[1 + L_b/2ESL_d]$. For low-$ESL_d$ (i.e., good) decoupling capacitors and layouts, the condition $e^{P|a|} >> e^{-P|a|}$ continues to be fulfilled. Therefore, the middle-segment model continues to prevail at higher frequencies for even as few as 3 DIPs. Similarly, the applicability of the end segment models is enhanced by low $ESL_d$.

The effects of departures from the conditions $e^{P|a|} >> e^{-P|a|}$ and $e^{K|a|} >> e^{-K|a|}$ are noted at frequencies which are high in comparison to the self-resonant frequency of the decoupling capacitor. The most important practical effect is deterioration of suppression of conducted noise.

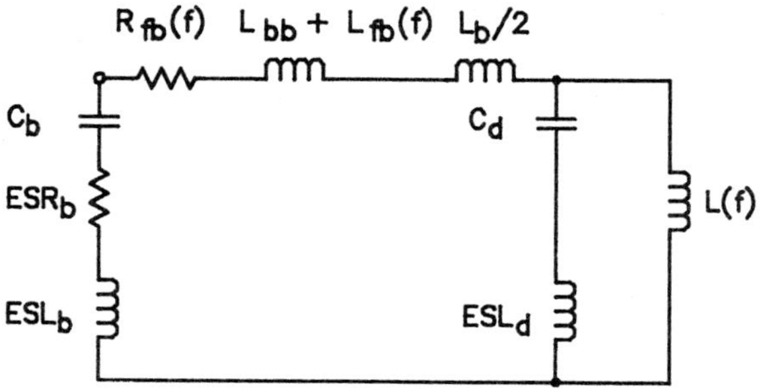


A. Power Entry/Feed and First Segment of Power Bus

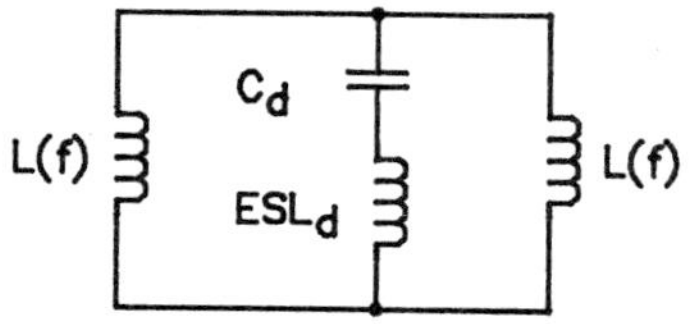


B. Middle Segments of Power Bus

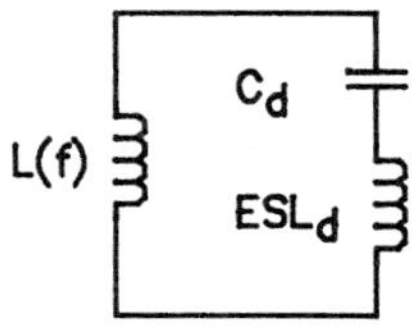


C. End Segment of Power Bus

Figure 3.3-1: HIGH-FREQUENCY MODELS OF SEGMENTS OF THE POWER BUS

**A Worst-Case High-Frequency Model**

The end-segment model shown in Figure 3.3-1C is the worst case for high frequencies, as the inductance is maximum for that segment. Therefore, the tendency to ring and generate noise will be maximum. Addition of the DIP equivalent circuit of Figure 3.2-2B to the model in Figure 3.3-1C leads to Figure 3.3-2A.* In that Figure, $L_d$ is the inductance incurred in connecting the gates to the decoupling capacitor [see (3.2-1) for a definition of $L_d$, and Figure 3.2-1 for an example of the traces and internal DIP wiring comprising that inductance].

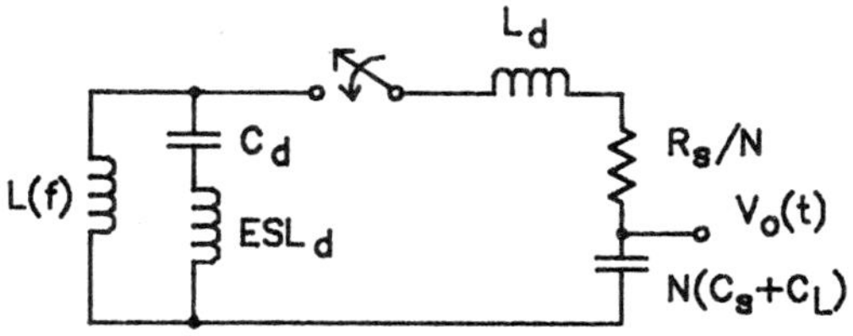


A. First Step Towards Derivation of Model

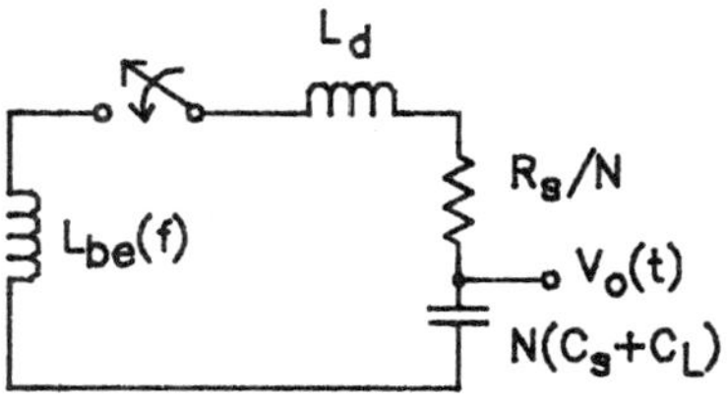


B. The Model

Figure 3.3-2: WORST-CASE HIGH-FREQUENCY MODEL OF A DISCHARGE LOOP

For $f > f_d$, the decoupling capacitor in Figure 3.3-2A is exactly equivalent to an inductor $L_c(f) = ESL_d[1 - (f_d/f)^2]$. Paralleling that inductor with the $L(f)$ in Figure 3.3-2A [and given in (3.3-4)] leads to the simplified

---

* In Figure 3.3-2, only the high-frequency components of the output, $v_o(t)$, are of concern. Batteries appearing in the DIP equivalent circuit of Figure 3.2-2B do not appear in the high-frequency circuits.

model in Figure 3.3-2B. The effective inductance due to the decoupling capacitor in parallel with the inductive impedance of the rest of the chain is

$$L_{be}(f) = \frac{[1 - (f_d/f)^2]\ ESL_d}{1 + \{ESL_d[1 - (f_d/f)^2]/L(f)\}} \tag{3.3-5}$$

$$\cong \frac{ESL_d}{1 + \dfrac{2ESL_d}{L_b[1 + (1 + 4ESL_d/L_b)^{1/2})}}, \qquad (f/f_d)^2 >> 1$$

To keep the high-frequency model relatively simple, neither the DIP-internal gate-to-output pin inductance nor the inductance of the trace connecting the output of the DIP to the (gate) load, $C_L$, are considered. In practice, this condition can be approximated by maintaining extremely short output traces. However, this is oftentimes difficult to do when other PCB layout constraints are considered. Neglect of inductances associated with the outputs is, then, a signifigant limitation of the present model. However, due to the relative isolation of the inductances associated with the N x F output traces by the output source resistance $R_S$, the effects of those inductances on the voltage across the decoupling capacitor, on emissions, and on integrities of other output pulses are, in practice, minimal to modest.

At the high frequencies defined in (3.3-1), **all segments of the K-DIP chain will perform better than the worst-case model given in Figure 3.3-2.**

**Low-Frequency Model**

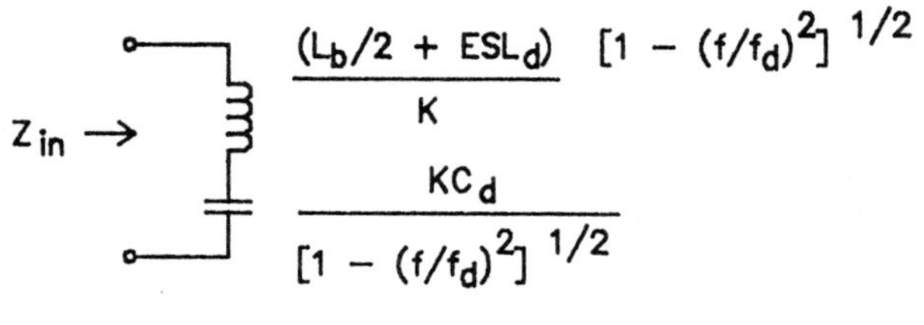


Fig. 3.3-3: LOW-FREQUENCY MODEL OF THE INPUT IMPEDANCE OF A K-SEGMENT CHAIN

At low frequencies defined in (3.3-2), concern is with the input impedance of the entire chain of segments beginning at the end of the power feed traces and ending at the last decoupling capacitor. The input impedance of the model shown in Figure 3.3-3 matches the expression for the low-frequency input impedance of the entire chain of segments derived in Appendix 1.

In practical designs, the inductive component of the input impedance at the power entry point is dominated by the inductance of the power feed traces, $L_{bb}$, and the inductance of the ferrite bead, $L_{fb}(f)$. The inductance in the Figure above is negligible in formulating an expression for the input impedance of the power bus as seen from the power entry point. The capacitance, approximately $KC_d$ at frequencies substantially below the self-resonant frequency of the decoupling capacitor, is not negligible.

**A Worst-Case Low-Frequency Model**

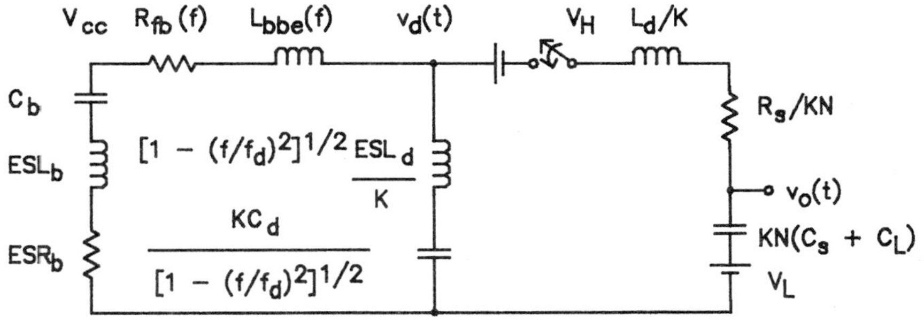


A. First Step in Derivation of Model

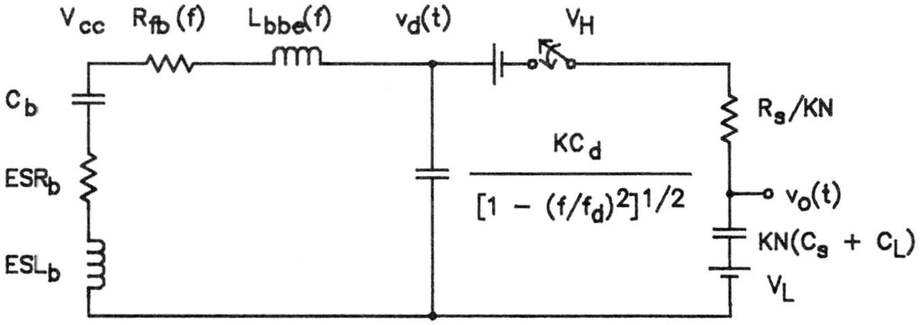


B. The Model

Figure 3.3-4: WORST-CASE LOW-FREQUENCY MODEL OF A CHARGING LOOP

The model in Figure 3.3-4A is similar to that given in Figure 3.3-1A with the effects of the switched gates in the DIPs added; the low frequency equivalent circuit in Figure 3.3-3 is incorporated. In the Figure, the effective inductance, $L_{bbe}(f)$, is the sum of the ferrite bead inductance, the inductance of the power feed traces, and the comparatively small inductance representing the input impedance of the K-segment chain shown in Figure 3.3-3:

(3.3-5)

$$L_{bbe}(f) = L_{fb}(f) + L_{bb} + L_b/2 + (\{L_b/2 + ESL_d\}[1 - (f/f_d)^2/K]^{1/2}\}$$

$$\cong L_{fb}(f) + L_{bb} + L_b/2$$

In all practical designs, the low frequencies defined in (3.3-2) are much less than the resonant frequency of the right-hand loop of Figure 3.3-4A. Therefore, the inductor $L_d$ in that loop may be neglected. Additionally, $ESL_d[1 - (f_d/f)^2]^{1/2}/K$ is neglible in comparison with $L_{bbe}(f)$. With these approximations, the resulting model is that given in Figure 3.3-4B.

The low-frequency noise generated due to the switch closure appears at all DIPs with approximately equal magnitude, as the frequencies are lower than those which are effectively filtered by the LC chain.

3.4 Susceptibility to Noise on the Power Bus

The aim of the next several Sections is to use the worst-case models developed in Section 3.3 to insure that the waveshape of the DIP output pulse results in noise-free operation of the driven gate(s). As a prelude, the emphasis of the present Section is on defining what is meant by pulse integrity and susceptibility to noise on the power bus. Emissions issues are addressed in Sections 3.8 and 3.9.

By "pulse integrity" is meant generation of near square-wave output pulses by DIP output gates. In particular, concern is with design of decoupling networks to provide freedom from

- Ringing of the power bus,
- Excessive rise time, and
- Preservation of DC levels at appropriate points in the power distribution network.

It can be seen from the equivalent circuit of a DIP in Figure 3.2-2 that the rise time of a gate can be affected by the decoupling network, but the fall time is not affected. Thus, concern here is with excessive rise time only.

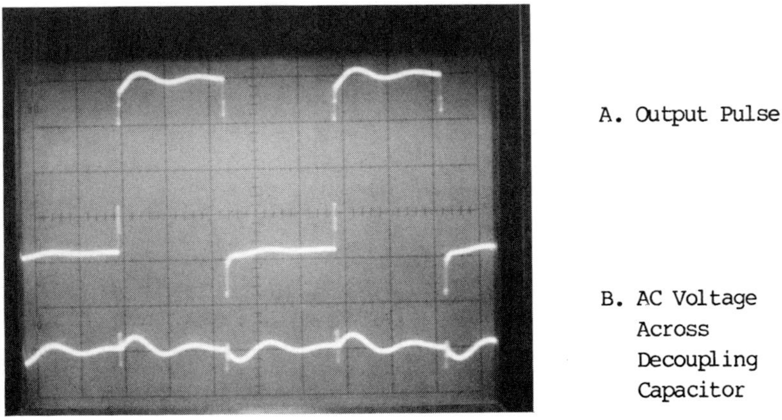


Vertical = 1 V/cm, Horiz = 1 us/cm. LSTTL referenced to power entry ground.

Figure 3.4-1: OUTPUT FROM A POORLY-DECOUPLED DIP

**Decoupling Influences Pulse Integrity**

Current drawn from the power bus during a gate's transition from low to high can cause ringing of the power bus. That noise is superimposed on the output signal(s) from a DIP, possibly causing the driven gate to misinterpret the noisy input signal. From a susceptibility standpoint, **the purpose of a decoupling network is to prevent ringing-type noise on the power bus.**

**Ground Inductance Also Influences Pulse Integrity**

Due to voltage drop across ground and DIP-internal inductances, a transition from high to low, as well as the aforementioned low-to-high transition, can cause ringing to be observed when, as in Figure 3.4-1, a gate's output is measured with respect to a ground other than the ground end of the decoupling capacitor serving that output gate (see Figure 3.2-6A for reasons).

Unfortunately, the driven gates' grounds are oftentimes not connected via low impedances (e.g., a ground plane, semi-ground plane* or grid [10]) to the ground end of the decoupling capacitor at the driving gate. Relatively large-amplitude ground noise results:

**Ground Noise** = Difference in voltage between the ground pin of the DIP in which the driving gate resides and the ground pin of the DIP in which the driven gate resides.

Figure 3.4-2 shows the case of an output gate driving a single input gate. Ground noise is the voltage developed across the ground trace inductance. That noise is impressed on the input of the driven gate.

**Decoupling, Not Ground Inductance, is of Primary Concern in This Book**

**Good decoupling is a necessary, but not sufficient, condition for achievement of minimum susceptibility and emissions.** In this book, the effects of ground and signal trace inductances - as incurred in distributing a gate output signal to an input gate - are analytically neglected, as in Figure 3.4-2B. Although ground inductance considerations are implicit in PCB layout (Section 4.1) and other recommendations and discussions, the emphasis in this book is on the decoupling problem, which does involve power bus and other trace inductances. Signal degradation due to the inductive/transmission-line effects of signal and ground traces is discussed in [1].

---

* On a two-sided PCB, a semi-ground plane is achieved by leaving as much copper as possible on the PCB and connecting the resulting "spare" lands together and to ground wherever possible. This practice is recommended.

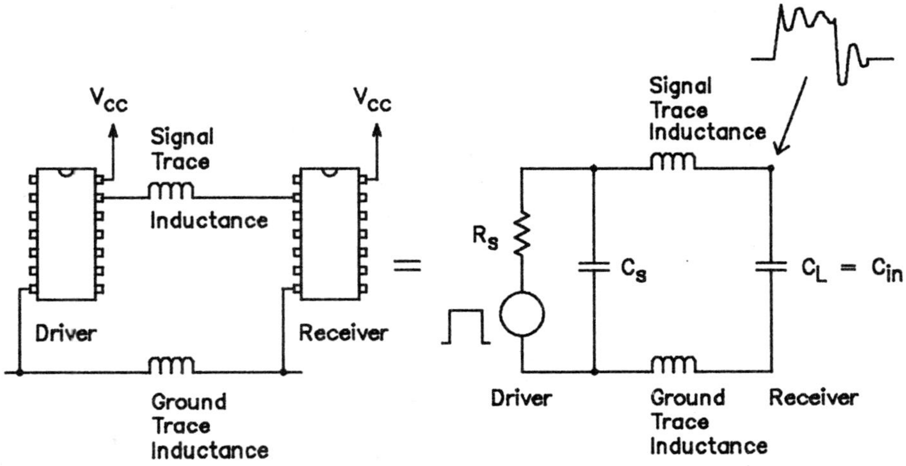


A. The Ground Inductance Problem

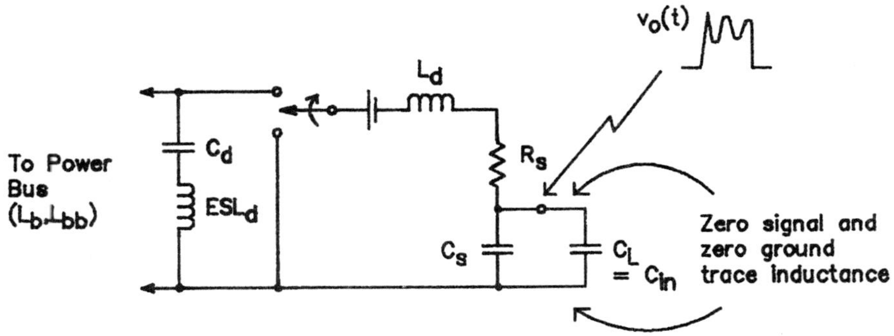


B. The Decoupling Problem

Figure 3.4-2: DISTINGUISHING BETWEEN GROUND NOISE AND DECOUPLING PROBLEMS

## Criteria for Noise Immunity

Pulse integrity is necessary in order to meet criteria for both "high" and "low" inputs to the driven gates. **A circuit's susceptibility to power bus disturbances - caused, e.g., by operation of other circuits connected to the same DC power bus - is inversely proportional to the conservatism with which noise immunity criteria are meant.**

The aforementioned criteria are cited in Table 3.4-1 for various technologies [4 through 7]. "Typical" noise immunity is larger. "Noise Immunity" is the difference between the gate output and input levels for a given state; e.g., the noise immunity for LSTTL is 0.8 - 0.5 = 300 mV for the low state and 2.7 - 2.0 = 700 mV for the high state, indicating a larger tolerance for power bus noise in the high state. Noise on the power bus, as measured from the output(s) to the ground pin on a DIP, must be such that the levels cited in the Table are not exceeded. Otherwise, erratic logic operations are possible.

Experience has shown that less than 50 mV, peak-to-peak (pp), noise voltage at the PCB power entry point is usually indicative of good decoupling practice for all but ECL technology, where 20 mV pp is appropriate. Less than 50 mV pp at the power entry point does not guarantee that the voltage fluctuation measured between the $V_{CC}$ and ground pins of all DIPs on a PCB will be less than the noise immunity level, but it is one "rule of thumb" measure used to obtain a general idea of the state of decoupling network design.

| | $V_{oL}$ | $V_{iL}$ | $V_{oH}$ | $V_{iH}$ | **Noise Immunity** Low/High |
|---|---|---|---|---|---|
| CMOS [B Series] | 0.05 | 0.3 $V_{CC}$ | $V_{CC}$ -.05 | 0.7$V_{CC}$ | ≅ 0.3$V_{CC}$ |
| HCMOS | 0.10 | 1.0 | 4.9 | 4.5 | 900 mV/400 mV |
| LSTTL | 0.50 | 0.80 | 2.7 | 2.0 | 300 mV/700 mV |
| ECL [10K, 10KH] | -1.63 | -1.48 | -0.98 | -1.1 | 150 mV/120 mV |

$V_{oL}$ = Maximum gate output voltage, low state

$V_{iL}$ = Maximum gate input voltage, low state

$V_{oH}$ = Minimum gate output voltage, high state

$V_{iH}$ = Minimum gate input voltage, high state

Table 3.4-1: CRITERIA FOR HIGH AND LOW OUTPUT PULSES

**Timing Constraints**

Avoidance of logical errors due to timing (race conditions) provides another criterion for pulse integrity. Propagation delay and the propensity for timing errors is affected by the rise and fall times with which gates achieve high and low states, respectively. In [1, p. 2.2-2], it is shown that the 10 % to 90 % rise time or the 90 % to 10 % fall time for a perfectly-decoupled circuit is given by

$$\tau = 2.20 \times [R_S(C_S + C_L)], \tag{3.4-1}$$

where $R_S$ and $C_S$ are the intrinsic source resistance and capacitance in Table (3.2-1). As will be shown, poor decoupling can increase rise time.

**A Paucity of Noise-Related Data**

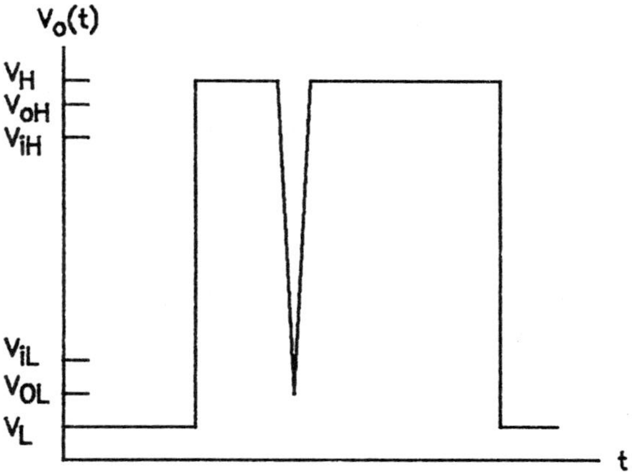


Figure 3.4-3: A GLITCH IN THE OUTPUT SIGNAL OF A DRIVING GATE

Although these data are not provided by any manufacturer of which the author is aware, the question arises as to how long an output pulse may be less than the minimum guaranteed input gate "on" level, $V_{iH}$, before a driven gate is "tricked" into thinking that an "off" pulse is present. In practice, it has been observed that, for an error to result in the output of the gate being driven by the signal in the Figure, the driving output gate voltage must be less than $V_{iH}$ for approximately the rise time of the driven gate. The probability of a glitch similar to that shown in the Figure in the output of the driven gate is greatly enhanced by the (driving) output pulse not only being less than $V_{iH}$, but by being less than $V_{iL}$. Between $V_{iH}$ and $V_{iL}$, the probability of an error occurring in the output of the (glitch-driven) gate increases with the duration of the glitch.

Unfortunately, there appears to be little data from manufacturers quantifying the above phenomenon. Data are important from the standpoint of assessing noise immunity to ringing of the output pulse and ground noise.

## 3.5 Background for Interpretation of Waveforms

Succeeding Sections provide quantitative analyses of the models given in Section 3.3. The purpose here is to provide an introduction to those analyses by interpreting key waveforms. For this purpose, it is helpful to note that the LRC loops in each of the worst-case models in Figures 3.3-2 and 3.3-4 must be overdamped to prevent ringing of the power bus. A review of fundamental damping notions is provided in Table 3.5-1.

During the course of interpretation of waveforms, it is helpful to provide a numerical example. The example cited in Table 3.5-2 is chosen as a good, but not excellent, design. Lead and trace inductances are given in (3.2-1), and component selection rationale is given in Section 4.2.

Figure 3.5-1 illustrates decoupling-related waveforms for the case of (proper) overdamping of the discharge loop illustrated in Figure 3.3-2B.*

**"Spikes" Across Decoupling Capacitors**

In Appendix 2, it is shown that the voltage spike across the decoupling capacitor at the instant of gate turn-on is

$$(V_H - V_L)F_2 = \frac{V_H - V_L}{1 + L_d/L_{be}(\infty)} \qquad (= 633 \text{ mV}) \qquad (3.5\text{-}1)$$

where, here, the practice of providing parenthetical "typical" results based on the example cited in Table 3.5-2 is begun. $L_{be}(\infty)$, 12.7 nH, is from (3.3-5) using element values given in (3.2-1).

This spike occurs due to the "argument" between capacitors, across which the voltage cannot be instantaneously changed, and inductors, through which the current cannot be instantaneously changed, when the gates are turned on.

---

* Arguments for the 50 % clock duty cycle usually encountered in practice and shown in the Figure are as follows: given that the gates can change state twice per clock cycle, maximum time is allowed for recovery of the charge of the decoupling capacitor between gate turn-ons. Also, from an emissions standpoint, even harmonics of an ideal clock waveform disappear when the clock duty cycle is 50 % (this is an interesting experiment with a pulse generator and spectrum analyzer: around a given center frequency, the even harmonics null out as the duty cycle of the pulse generator nears 50 %, but, due to a non-ideal pulse, the center frequency around which nulling occurs is very sensitive to the duty cycle setting of the pulse generator).

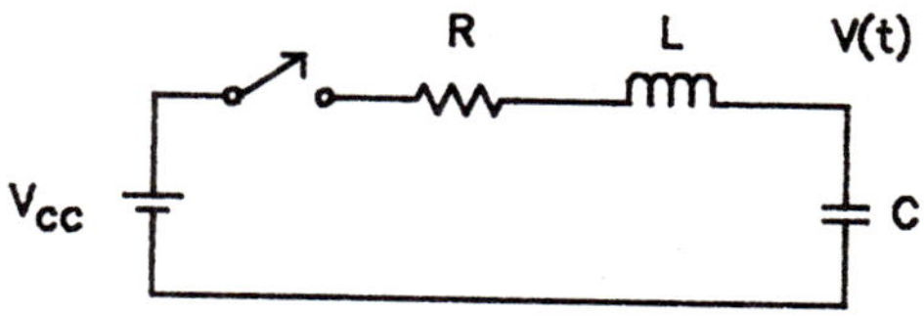


If $D = \frac{4L}{R^2C} > 1$, the circuit is UNDERDAMPED and rings:

$$v(t) = V_{cc}[1 - e^{-at}(\cos Bt + (a/B)\sin Bt)], \text{ where}$$

$$a = R/2L \qquad \text{and} \qquad B = |1/LC - a^2|^{1/2}$$

If $D = \frac{4L}{R^2C} < 1$, the circuit is OVERDAMPED and does not ring:

$$v(t) = V_{cc}[1 - e^{-at}(\cosh Bt + (a/B)\sinh Bt)], \text{where}$$

a and B are as given above. If $D << 1$ (strong overdamping),

$$v(t) \cong V_{cc}[1 - e^{-t/RC}], \ t >> 1/RC$$

Table 3.5-1: REVIEW OF DAMPING FOR A SERIES RLC CIRCUIT

Technology: LSTTL

Clock Frequency: 4 MHz.

Layout: Figure 3.2-1 (Detailed trace widths, etc., given in Appendix 3).

Power Entry Capacitor: $C_b$ = 4.7 uF, $ESL_b$ = 10 nH, $ESR_b$ = 0.2 Ohms.

Ferrite Bead: See Figure 3.2-3; > 40 Ohms resistance for f > 10 MHz.

Decoupling Capacitors (6): $C_d$ = "0.01" uF (measured at 0.007 uF - see Appendix 3), $ESL_d$ = 24.8 nH, 6.8 nH of which is due to trace and lead inductance, $ESR_d$ = 0.2 Ohms.

Fanout and Load: N = 5 Gates/DIP going on with each gate loaded with $C_L$ = 10 pF, representing a fanout of two (F = 2) to LSTTL input gates.

Table 3.5-2: ELEMENT VALUES FOR EXAMPLE

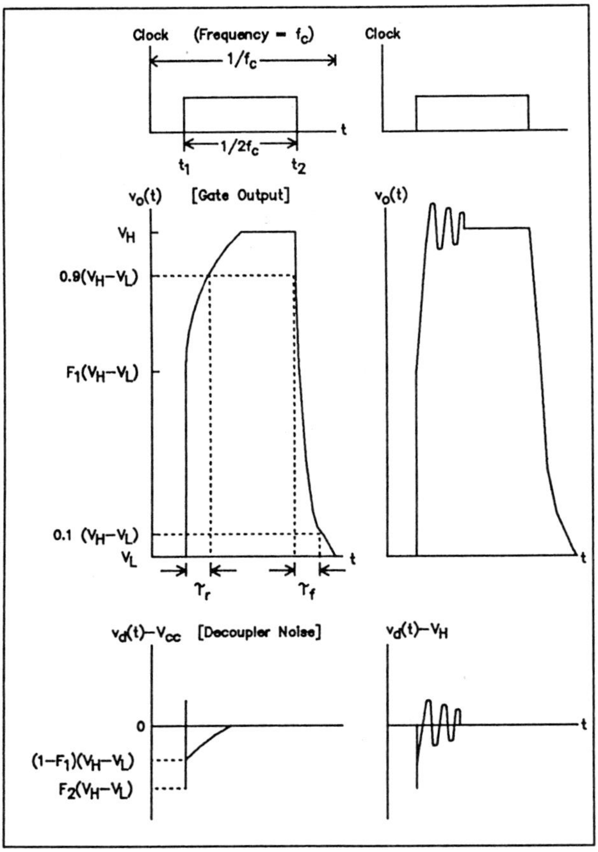


A. Overdamped Charging Loop B. Underdamped Charging Loop

Figure 3.5-1: EXAGGERATED FEATURES OF GATE AND DECOUPLING CAPACITOR VOLTAGES

Proper overdamping of the discharge (not charging) loop is depicted in both Figures 3.5-1A and B. Then, for both the underdamped and overdamped charging loop cases, the duration of the spike is small, as this is quickly damped out in the discharge loop after a brief excursion.

**Output Pulse**

The spike which occurs across the decoupling capacitor also occurs at the gate output, but it is not apparent because it is subsumed in the transition interval and never causes a negative-going output. Then, for either the underdamped or overdamped conditions of Figure 3.5-1, when the N gates turn on at $t = t_1$, the gate output voltage, $v_o(t)$, rapidly rises from $V_L$ to $F_1(V_H - V_L)$. By virtue of the inductances in the circuit - which, in effect, "tune out" part of the C in the RC network surrounding the gates - this rise takes place more rapidly than is determined via the "pure" RC expression (3.4-1).

After the spike and fast rise, the initial voltage at the gate output is due to the redistribution of total charge between the decoupling capacitor and the output gate and load capacitance:

$$F_1(V_H - V_L) = \frac{V_H - V_L}{1 + N(C_s + C_L)/C_d} \qquad (3.5\text{-}2)$$

(= 3.23 V; absolute level = 3.23 V + 0.2 V = 3.43 V).

This gate output voltage corresponds to a commensurate decrease in the voltage across the decoupling capacitor which takes place so rapidly that the relatively "sluggish" charging loop cannot recharge the decoupling capacitor during the gate transition. The charging loop is, in effect, disconnected during the gate transition interval.

The remainder of the voltage rise to $V_H$ (to 3.5 V) occurs over a time period established by the charging loop. The manner in which this rise is achieved depends on the damping of that loop. This is discussed in detail in subsequent sections.

**A Lower Bound on the Capacitance of the Decoupling Capacitor**

From Figure 3.5-1, it is apparent that if $F_1(V_H - V_L)$ is greater than $0.9(V_H - V_L)$, the rise time of the circuit, as measured from the 10 % to 90 % points, is not increased by the decoupling network (indeed, the inductances serve to decrease the 10 % to 90 % rise time). By requiring that $F_1 (V_H - V_L) \geq 0.9 (V_H - V_L)$ in (3.5-2),

$$C_d \geq 9N(C_s + C_L), \qquad (= 0.00135\ \text{uF}), \qquad (3.5\text{-}3)$$

insures that - even without inductances - the rise time is not increased by the decoupling network. (3.5-3) is adopted as a design guideline.

## 3.6 Damping of the Charging Loop

In good designs, it is necessary to overdamp both the charging and discharging loops illustrated in Figures 3.3-4 and 3.3-2, respectively. This Section is concerned with damping of the charging loop.

### Analysis of the Charging Loop at Low Frequencies

Figure 3.6-1 illustrates the worst-case model in Figure 3.3-4B. The element values are for the example in Table 3.5-2 with ferrite bead characteristics illustrated in Figure 3.2-5. The numerical values for $L_{fb}$(4 MHz) and $R_{fb}$(4 MHz) are shown (this selection of frequency is discussed below).

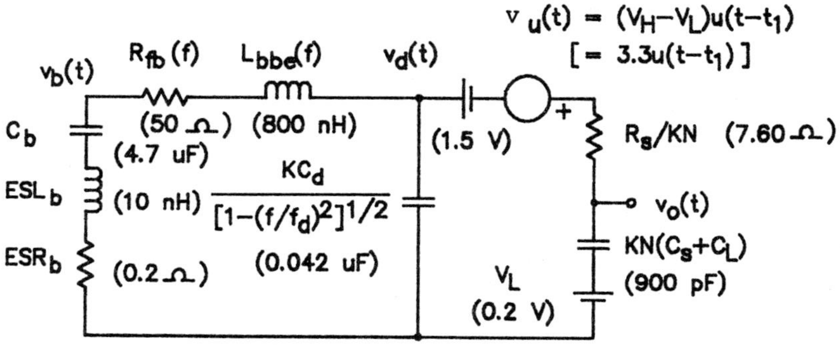


Values shown are for LSTTL, K = 6, N = 5, $C_L$ = 10 pF (Fanout = 2/Gate).

Figure 3.6-1: WORST-CASE LOW-FREQUENCY DECOUPLING

In the Figure, the technique of adding a unit step generator, $u(t-t_1)$, actuating when the N gates turn on at $t = t_1$, is a standard one for analysis of transients in switch-closure circuitry [8, pp. 51-56]. $v_b(t)$, and $v_d(t)$ are the steady-state level ($V_{CC}$) plus the voltages induced by the step generator. The assumption of a steady-state level implies that preceding transients have, for practical purposes, evanesced during one-half the clock period, $1/2f_c$, where

$$f_c = \text{Clock frequency} \qquad (= 4 \text{ MHz}). \qquad (3.6\text{-}1)$$

Methods for insuring that the transients die out during one-half the clock period are outlined below.

The model is valid for low frequencies: $f < f_d/[1 + L_b/4ESL_d]^{1/2}$ (= 11.3 MHz). Concern is with damping of the left-hand (charging) loop to prevent possible ringing, the frequency of which <u>is</u> within that low-frequency range.

The conditions

$$C_b \gg KC_d \qquad (4.7\ \text{uF} \gg 0.042\ \text{uF}), \qquad (3.6\text{-}2)$$

$$C_d \gg N(C_s + C_L) \qquad (0.007\ \text{uF} \gg 150\ \text{pF}),$$

prevail in practice. Provided that $C_d \geq 9N(C_s + C_L)$ [per (3.5-3)], the reactance of the capacitor $KN(C_s + C_L)$ is at least nine times larger than that of of the capacitor $KC_d$. In usual practice and in the example, the ratio is larger. Then, the resistance $R_s/KN$ does not significantly contribute to damping of the left-hand (charging) loop, and damping may be determined without regard to that resistance.

Under the above conditions, it follows from applying the expression for damping given in Table 3.5-1 to the left-hand loop of Figure 3.6-1 that

$$D_b = \frac{4(L_{bbe}(f) + ESL_b)\,[1 - (f/f_d)^2]^{1/2}}{(R_{fb}(f) + ESR_b)^2 KC_d} < 1 \qquad (= 2.88 \times 10^{-2}\ @\ 4\ \text{MHz}) \qquad (3.6\text{-}3)$$

is required for overdamping of the charging loop at f = 4 MHz. As discussed below, selection of an appropriate ferrite bead insures that this damping requirement is met. **A properly-selected ferrite bead stops charging loop ringing.** Good design for pulse integrity and low emissions leads to circuit element values which would otherwise cause low-frequency ringing.

Stating (3.6-3) another way, the minimum value of $C_d$ required to assure proper overdamping ($D_b < 1$) is

$$C_d > \frac{4(L_{bbe}(f) + ESL_b)\,[1 - (f/f_d)^2]^{1/2}}{K(R_{fb}(f) + ESR_b)^2} = \begin{cases} 200\ \text{pF with ferrite bead,} \\ 0.158\ \text{uF without ferrite bead} \\ \{L_{fb}(f) = R_{fb}(f) = 0\}, \end{cases} \qquad (3.6\text{-}4)$$

where the numerical data are for the example.

**Rise time considerations stated in (3.5-3) determine the minimum value of $C_d$ when a properly-selected ferrite bead is employed: $C_d \geq 9N(C_s + C_L)$. Otherwise, damping considerations (3.6-4) determine the (larger and more expensive) minimum value of $C_d$.**

**Damping at Frequencies Other Than Those Near the Clock Frequency**

Damping at frequencies substantially below the clock fundamental frequency are of little concern, as the amplitude excursions caused by ringing at those frequencies occurs over a small part of the ringing cycle and, therefore, are small.

At frequencies above the clock frequency (but below the frequency at which the low-frequency model in Figure 3.6-1 ceases to be valid), ringing <u>is</u> of concern. In usual practice, $L_{bbe}(f) \cong L_{fb}(f)$ and $R_{fb}(f) >> ESR_b$ in (3.6-3) and (3.6-4).

For most ferrite beads employed in equipment where the clock frequency is 1 MHz or higher, even a "randomly" selected ferrite bead exhibits element values, $L_{fb}(f)$, $R_{fb}(f)$, which are changing with increased frequency in such a way to lower $D_b$ (i.e., increase damping). Additionally, the factor $[1 - (f/f_d)^2]^{1/2}$ in (3.6-3) increases damping (decreases $D_b$) as the frequency is increased beyond the clock frequency. As is easily verified from the characteristics of the ferrite bead given in Figure 3.2-5 and the expression (3.6-3), damping at frequencies above the clock frequency up to the limit of the model (11.3 MHz) is greater than that at the clock frequency.

For arbitrary clock frequencies, part of the ferrite bead selection process is to insure overdamping at the clock frequency and at all frequencies between the clock frequency and the upper-frequency limit defined in (3.3-2).

**Alternative Approaches to Damping**

Reference here is to damping with no ferrite bead. Then, from (3.3-5), $L_{bbe}(f) = L_{bb} + L_b/2$ and $R_{fb}(f) = 0$ in (3.6-3) and (3.6-4).

One approach to reducing the value of the required decoupling capacitor is to lay out the PCB to minimize the inductances of the power feed and inter-DIP power bus traces, $L_{bb} + L_b/2$. However, impractical PCB layout problems can occur and, as will be shown in Section 3.9, modest to moderate $L_{bb} + L_b$ is desirable from the standpoint of aiding filtering of high-frequency conducted noise.

Attempts to reduce the value of $C_d$ required for overdamping by allowing the equivalent series resistance of the power-entry capacitor, $ESR_b$, to increase has the undesirable effect of decreasing the suppression of conducted emissions at all frequencies (discussed in Section 3.9).

**Advantages and Selection of Ferrite Beads for Decoupling**

Using a properly selected ferrite bead at the power entry point provides for

- Lower-value, physically smaller and less expensive decoupling capacitors,
- Easier PCB layout (more inductance is allowed), and
- Better emissions performance (shown in Sections 3.8 through 3.10).

From the preceding, it is apparent that one important aspect of good decoupling design is selecting a "proper" ferrite bead. The discussions below are intended to provide selection criteria by discussing effects which can occur when the bead is imperfect in its operation.

Selection of an adequately large value of decoupling capacitor per (3.5-3) insures that the rise time of the output gates is not increased by the decoupling network, regardless of the ferrite bead selected. The discussions below pertain to ringing and to how fast the last 10 % of the rise is achieved after the 10 % to 90 % rise has been achieved.

A tradeoff between larger-valued decoupling capacitors and bead characteristics can be performed by using the characteristics of the bead, $L_{fb}(f)$ [part of $L_{bbe}(f)$] and $R_{fb}(f)$, in the expression (3.6-4). In practice, it is relatively straightforward to select a bead which concurrently insures overdamping and a value of decoupling capacitance which is substantially below that which would be required if the bead was not present {see, e.g., the numerical results in (3.6-4)}.

At the extreme of ferrite bead/decoupling capcitor selection possibilities is excessive overdamping. Then, particularly for high clock frequencies, the last 10 % of the rise of the output pulse to $V_H$ may not be achieved during one-half a clock cycle.

From Table 3.5-1 for the case of large overdamping, the requirement that the voltage recover to within 1 % of $V_H$ over one-half a clock cycle leads to the restriction

$$R_{fb}(f) + ESR_b < \frac{[1 - (f/f_d)^2]^{1/2}}{2f_c K C_d \ln[100z/(1 + z)]}, \quad z = N(C_s + C_d)/C_d, \qquad (3.6\text{-}5)$$

(< 3.79 Ohms for $f_c$ = 4 MHz, < 37.9 Ohms for $f_c$ = 0.4 MHz).

Thus, for large overdamping ($D_b$ << 1), the upper bound on low-frequency bead resistance is inversely proportional to clock frequency. Higher clock frequencies suggest use of beads which have lesser resistance at those clock frequencies. This allows full recovery of the decoupling capacitors' charge during one-half clock cycle, but, if that lesser resistance also characterizes the bead at high frequencies, suppression of conducted noise suffers (see Section 3.9).

**Example**

Figure 3.6-2 illustrates an output pulse from a circuit similar to that of the example. Similarities include use of a 0.007 uF decoupling capacitor and a ferrite bead with characteristics shown in Figure 3.2-5. Damping of the charging loop is adequate.*

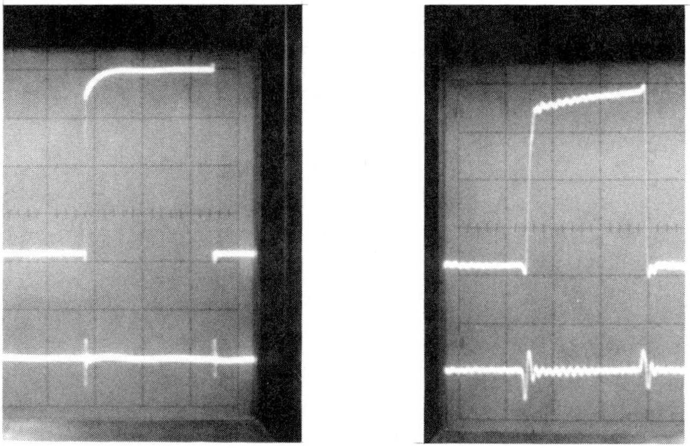

Gate Output

Voltage Across Decoupling Cap (Scope AC Coupled)

$f_c \cong 0.4$ MHz, 1 us/cm $\qquad$ $f_c \cong 4$ MHz, 100 ns/cm

Vertical = 1 V/cm. Same LSTTL circuit used for both photos.

Figure 3.6-2: OUTPUT PULSE FROM A CIRCUIT SIMILAR TO THE EXAMPLE

---

* The slight high-frequency ringing shown in the right-hand photo occurs in the discharge loop (see Section 3.7). The spike across the decoupling capacitor for a high-to-low transition occurs because the scope ground was not connected directly to the ground pin of the DIP.

Since, for the example, the bead resistance is 50 Ohms at $f_c$ = 4 MHz, the restriction (3.6-5) is not met at $f_c$ = 4MHz. The "tilt" in the waveform shown in the right-hand side of Figure 3.6-2 is predicted by the above analysis. However, that waveform is satisfactory for most purposes, and, even, desirable from an emissions standpoint insofar as rounding of the leading edge of the pulse reduces the high-frequency content.

Much of the "tilt" in the right-hand side of Figure 3.6-2 would be removed had the minimum value of the decoupling capacitor dictated by rise time considerations [0.00135 uF; see (3.5-3)] been used. Then, (3.6-5) leads to $R_{fb}$(4 MHz) < 19.7 Ohms.

It is better to err on the side of overdamping than underdamping, but limits are set by the desire to achieve the full noise immunity implied by maximizing the output swing over the entire duration of the pulse.

Ferrite bead selection recommendations are summarized in Section 4.2.

## 3.7 Damping the Discharge Loop

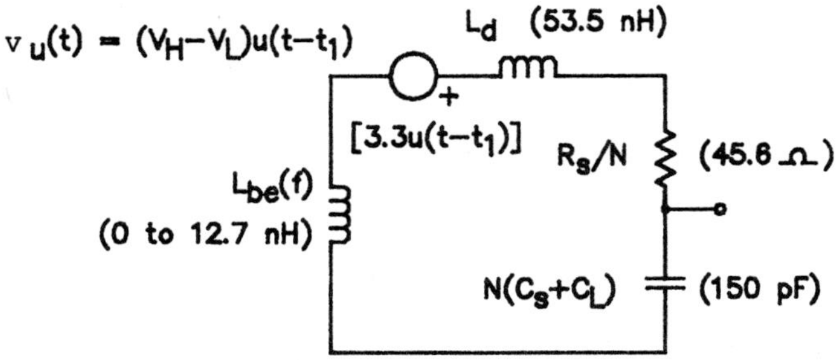


Values shown are for LSTTL, K = 6, N = 5, $C_L$ = 10 pF (Fanout = 2/Gate).

Figure 3.7-1: WORST-CASE HF/VHF DECOUPLING - THE DISCHARGE LOOP

Figure 3.7-1 illustrates the worst-case model shown in Figure 3.3-2B. Numerical data are for the example of Table 3.5-2. The ferrite bead does not appear in this model and has no effect on ringing of the discharge loop. The range for $L_{be}(f)$ given in the Figure is from (3.3-5) with f = $f_d$ to f = infinity.

Concern is with possible ringing, the frequency of which is well within the high-frequency range

$$f > \frac{f_d}{(1 + L_b/4ESL_d)^{1/2}} \qquad (f > 11.3\ \text{MHz}) \qquad (3.7\text{-}1)$$

Requiring that D in Table 3.5-1 be less than unity (D < 1) for the circuit in Figure 3.7-1 results in the following bound on the maximum number of gates which can be going on at once:

$$N < \frac{R_s^2(C_s + C_L)}{4[L_{be}(f) + L_d]} \qquad (N < 6) \qquad (3.7\text{-}2)$$

This restriction insures that the discharge loop is overdamped.

Practical element values are such that, if the restriction (3.7-2) is not met, ringing occurs. That ringing is always in the range (3.7-1). Practical element values are such that ringing is usually at VHF (30 MHz to 300 MHz).

To the extent that $C_d \gg N(C_s + C_L)$, **the result (3.7-2) is substantially independent of the value of the decoupling capacitor, $C_d$.** If $C_d \gg N(C_s + C_L)$ is already fulfilled, ringing of the discharge loop cannot be eliminated by selecting a larger-valued decoupling capacitor. Because $R_s$ and $C_s$ are determined by the logic family, the only cure for a fixed number of ongoing gates, N, is maintenance of low trace and decoupling capacitor inductance and, if the increased transition times can be tolerated, increasing the gate loads, $C_L$.

It is shown in Section 3.10 that the restriction (3.7-2) leads to N = 0 for ECL and HCMOS. Those families rely upon the dissipation of pull-up/pull-down or terminating resistors to achieve damping of the discharge loop, which the present model does not accommodate.

Extensions of the above results to layouts where multiple DIPs share a single decoupling capacitor are considered in Section 3.11.

## 3.8 Radiated Emissions

**Radiated emissions** are those received by an antenna. The primary purposes of this Section are to provide an introduction to radiated emissions problems due to inadequate decoupling and to place those problems in the perspective of other radiators typifying digital equipment.

Figure 3.8-1 is a typical spectrum received by an antenna near a computer. The three predominant peaks are harmonics of the clock (the absence of some clock harmonics is explained in [1, Section 3.6]). Concern is with clock-related noise on the power bus - the peaks - rather than the random signal-related noise manifested between the peaks in Figure 3.8-1.

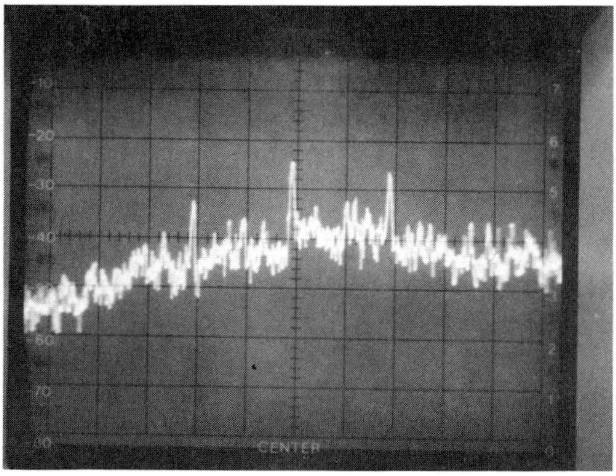


Horiz: 60 MHz cf, 2 MHz/cm. Vert: 0 dBuV baseline, 10 dB/cm (120 KHz BW)

Figure 3.8-1: RADIATED EMISSIONS FROM A 4 MHz-CLOCKED COMPUTER (TKC Photo)

### Radiated Emissions Due to Inadequate Decoupling

There are at least four ways in which inadequate decoupling can cause radiated emissions:

- **Direct Loop Radiation.** PCB power bus traces carry HF/VHF current at amplitudes determined in part by decoupling networks. The $V_{CC}$ trace and ground return trace or plane form a loop antenna, with radiation directly proportional to the product of that HF/VHF current, the square of the frequency* and the loop area between the traces, A. Also, free-space far-field radiation is inversely proportional to the distance, D, from the source of radiation to the receiving antenna:

$$\text{Maximum E-field radiation} = E = \text{Constant} \times \frac{I(f)\, f^2\, A}{D} \quad (\text{uV/meter}). \tag{3.8-1}$$

- **Loop Radiation from a Noisy Backplane/Motherboard.** PCB-generated power bus noise is conducted to the backplane or motherboard. Due to the larger loop areas involved, radiation from the backplane can exceed that from the PCB itself.

- **Conducted Noise to Output Ports.** Clock-related noise on the power bus serving the I/O DIPs appears at the output gates of those DIPs and then at the I/O ports of the equipment (at, e.g., an RS-232C interface). This results in common-mode noise current on cables connected to those ports, which causes radiation. Even for a cable with perfect shielding, noise will be radiated from an unshielded peripheral device connected to a noisy output port via that cable.

- **Conducted Noise to the Power Cord.** Clock-related HF/VHF noise on the $V_{CC}$ bus is transmitted back to the power supply serving the equipment. Most power supplies, and some power line filters, are transparent to that noise. Therefore, noise appears on the power cord, usually in the form of common-mode current at clock-harmonic frequencies. The power cord becomes an antenna which is quite efficient due to its length. Consequently, the equipment may pass conducted emissions testing, but can fail high-frequency radiated emissions testing due to radiation from the power cord at frequencies above those examined in testing for conducted emissions.

The method of Parker [11] concerned with radiation due to common-mode current in cables is useful in quantifying radiation due to conducted emissions at output ports and at the power cord. The previous work of the present author is useful in quantifying loop-based radiation stemming from differential-mode current in the power bus PCB traces and backplane [1, Sections 3.6 and 3.7].

---

* $f^2$ dependence applies for small PCBs. Loop perimeters approaching twice the four to eight inch lengths derived in (3.2-7) result in radiation dependency proceeding from $f^2 \rightarrow f \rightarrow f^{1/2}$ as f is increased beyond 300 MHz.

That part of PCB radiation of concern in decoupling is due to the loop formed by the power bus traces. That loop is characterized by relatively large differential-mode currents and physically small radiators (whereas cable radiation is characterized by relatively small common-mode currents and physically large radiators).

Which source of radiation - loop or cable - predominates in a given case depends on design details. For example, when metallic cabinetry with small apertures is used, cable radiation tends to predominate. With unshielded cabinetry, loop-based radiation from PCBs and backplanes can predominate (see, e.g., [1, pp. 3.6-1 - 3.7-6]). Even in the shielded cabinetry case, loop-based radiation can dominate the cause of system radiation by intra-cabinet coupling to internal cables and I/O traces/PCB's.

**A Hierarchy of Loop Radiators**

To put the problem of radiation from the power bus in perspective, consider the worst-case high-frequency model of Figure 3.3-2 for the practical case where the outputs are routed to relatively remote DIP input gates:

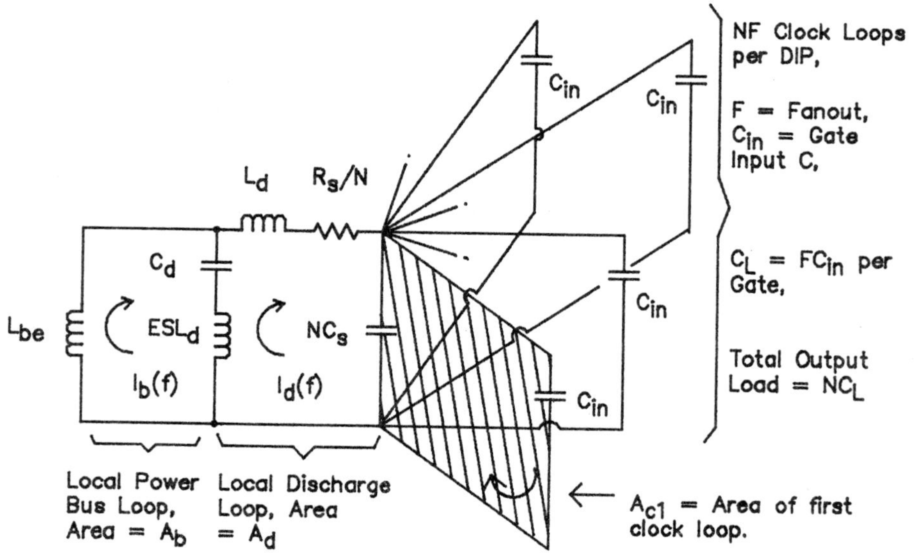


Figure 3.8-2: CLOCK LOOPS ASSOCIATED WITH A SINGLE DIP

In good designs, radiation from the power bus, including that from

- The traces defining the local discharge loop with area $A_d$,
- The traces comprising the local power bus loop, $A_b$, and
- The traces from the power-entry capacitor to the first DIP,

is less than that from the clock loops. Then, radiation from the power bus is less than from the clock traces. Once the clock traces become the dominant radiators, further suppression of radiation from the power bus should, in the interests of economics and common sense, stop.

During PCB layout, it is straightforward to confine the power bus and local discharge loops to relatively small areas. In contrast, fanouts associated with each of the N outputs must go to the input gates of relatively physically remote DIPs. The resulting loop areas formed by those clock traces and the associated ground plane(s) or traces is comparatively large. From (3.8-1), large loop areas lead to large radiation; in most cases, clock loops are the dominant radiators on PCBs [1].

However, due to the larger current circulating in the local discharge loop in Figure 3.8-2, radiation from the local discharge loop is a potential dominant radiator if the loop area, $A_d$, is not controlled. By inspection of Figure 3.8-2, the current in the local discharge loop is $(NC_s/C_{in})$ times that in a single single clock loop. Then, in order to keep the radiation from the discharge loop to a level which is less than that from the N x F clock loops connected to a single DIP, it follows from (3.8-1) that, for the case of equal clock loop areas and equal loads,

$$A_d \frac{NC_s}{C_{in}} < NF\,A_c, \text{ or} \qquad (3.8\text{-}2)$$

$$A_d < \frac{NFC_{in}}{NC_s} A_c = \frac{C_L}{C_s} A_c \qquad (A_d < 0.5\,A_c),$$

is required in order to maintain the radiation from a single local decoupling loop at a level which is less than that from the N output loops. This condition is usually "naturally" met in PCB layout. For example, it is shown in Appendix 3 that $A_d = 1.69\ cm^2$ for the layout of Figure 3.2-1. Due to the need to run the ouputs to other DIPs, it would be very difficult to provide an average clock loop area less than twice that value; i.e., the condition $A_d < 0.5A_c$ is met with ease.

Current in power feed traces is primarily at low frequencies, where the efficiency of the traces as antennas is greatly reduced. The preceding high-frequency considerations are much more important.

**Quantification of Radiated Emissions**

Table 3.8-1, useful for estimating radiation from power buses, clock traces, etc., is from [1, p. 3.2-9]. It is a quantification of free-space radiation which includes both (3.8-1) and near-field effects. Reflection of energy from ground can add up to 6 dB to the radiation levels given in the Table.

"Short-circuit cutoff perimeter" is the total trace length ("hot" plus return trace) at which the dependency of the E-field begins to decrease from $f^2$ to $f^v$, $v < 2$. For currents and loop areas greater than 1, add (in dB) those currents/loop areas to the radiation indicated in the Table; e.g., radiation from a 10 $cm^2$ loop carrying 0.5 mA of current at f = 100 MHz, as observed at D = 3 meters, is 13.0 + 20 log(10) + 20 log(0.5) = 27.0 dB uV/m (FCC limits are 40 to 56.9 dB uV/m).

At 300 MHz, radiation will begin to be less than that predicted (41.6 dB at D = 1 meter) for 1 mA in a 1 $cm^2$ loop when the total length of both traces approaches 28.87 cm (i.e., a trace-pair run of 14.44 cm ≅ 5.69") for a trace structure with an effective relative permittivity ($e_r'$) of 3.0. As discussed in Appendix 4, $e_r'$ depends on the permittivity of the PCB material and trace-run details (traces on same or opposite sides of the PCB, etc.).

Both loop-based radiation cited in Table 3.8-1 and "whip antenna"-based (i.e., cable with common-mode current-based) radiation contribute to the overall radiated emissions spectrum for typical equipment. **For all but extremely poor power bus layouts, loop-based clock-trace radiation is greater than loop-based radiation from the power bus.** Depending on design details, loop-based radiation from clock traces may or may not predominate whip antenna radiation. Whip antenna radiation is usually due to (conducted) power bus clock noise finding its way to an I/O cable or power cord, thereby causing common-mode noise current to flow in that cable or cord.

Section 3.9 following quantifies conducted noise, but does not carry the analysis to the next step of determining cable-based radiation due to that conducted noise becoming a common-mode current on a cable. Parker [11] has provided an analysis quantifying radiation once the more difficult task of determining the common-mode current is completed.

| f (MHz) | E (dB uV/m), 1 mA in 1 $cm^2$ Loop @ Observation Distance D (m) = 1 | 3 | 10 | 30 | 100 | Short-Circuit Cutoff Perimeter @ er' = 1.0 | 3.0 | 6.0 |
|---|---|---|---|---|---|---|---|---|
| 0.01 | -64.0 | -83.1 | -104.0 | -123.1 | -144.0 | 15.00 km | 8.66 km | 6.12 km |
| 0.02 | -58.0 | -77.1 | -98.0 | -117.1 | -138.0 | 7.50 km | 4.33 km | 3.06 km |
| 0.03 | -54.5 | -73.6 | -94.5 | -113.6 | -134.5 | 5.00 km | 2.89 km | 2.04 km |
| 0.04 | -52.0 | -71.1 | -92.0 | -111.1 | -132.0 | 3.75 km | 2.17 km | 1.53 km |
| 0.05 | -50.1 | -69.1 | -90.1 | -109.1 | -130.0 | 3.00 km | 1.73 km | 1.22 km |
| 0.06 | -48.5 | -67.6 | -88.5 | -107.6 | -128.4 | 2.50 km | 1.44 km | 1.02 km |
| 0.07 | -47.1 | -66.2 | -87.1 | -106.2 | -127.0 | 2.14 km | 1.24 km | 0.87 km |
| 0.08 | -46.0 | -65.1 | -86.0 | -105.0 | -125.9 | 1.88 km | 1.08 km | 0.77 km |
| 0.09 | -45.0 | -64.0 | -84.9 | -104.0 | -124.8 | 1.67 km | 0.96 km | 0.68 km |
| 0.10 | -44.0 | -63.1 | -84.0 | -103.1 | -123.8 | 1.50 km | 0.87 km | 0.61 km |
| 0.20 | -38.0 | -57.1 | -78.0 | -97.0 | -117.3 | 0.75 km | 0.43 km | 0.31 km |
| 0.30 | -34.5 | -53.6 | -74.5 | -93.4 | -113.0 | 0.50 km | 0.29 km | 0.20 km |
| 0.40 | -32.0 | -51.1 | -72.0 | -90.8 | -109.7 | 0.38 km | 0.22 km | 0.15 km |
| 0.50 | -30.1 | -49.1 | -70.0 | -88.7 | -106.8 | 0.30 km | 0.17 km | 0.12 km |
| 0.60 | -28.5 | -47.6 | -68.4 | -87.0 | -104.4 | 0.25 km | 0.14 km | 0.10 km |
| 0.70 | -27.1 | -46.2 | -67.0 | -85.5 | -102.2 | 0.21 km | 0.12 km | 87.48 m |
| 0.80 | -26.0 | -45.0 | -65.9 | -84.1 | -100.2 | 0.19 km | 0.11 km | 76.55 m |
| 0.90 | -24.9 | -44.0 | -64.8 | -82.8 | -98.4 | 0.17 km | 96.23 m | 68.04 m |
| 1.00 | -24.0 | -43.1 | -63.8 | -81.7 | -96.7 | 0.15 km | 86.60 m | 61.24 m |
| 2.00 | -18.0 | -37.0 | -57.3 | -73.0 | -85.3 | 75.00 m | 43.30 m | 30.62 m |
| 3.00 | -14.5 | -33.4 | -53.0 | -67.0 | -78.4 | 50.00 m | 28.87 m | 20.41 m |
| 4.00 | -12.0 | -30.8 | -49.7 | -62.4 | -73.5 | 37.50 m | 21.65 m | 15.31 m |
| 5.00 | -10.0 | -28.7 | -46.8 | -58.8 | -69.6 | 30.00 m | 17.32 m | 12.25 m |
| 6.00 | -8.4 | -27.0 | -44.4 | -55.7 | -66.5 | 25.00 m | 14.43 m | 10.21 m |
| 7.00 | -7.0 | -25.5 | -42.2 | -53.1 | -63.8 | 21.43 m | 12.37 m | 8.75 m |
| 8.00 | -5.9 | -24.1 | -40.2 | -50.9 | -61.5 | 18.75 m | 10.83 m | 7.65 m |
| 9.00 | -4.8 | -22.8 | -38.4 | -48.9 | -59.4 | 16.67 m | 9.62 m | 6.80 m |
| 10.00 | -3.8 | -21.7 | -36.7 | -47.0 | -57.6 | 15.00 m | 8.66 m | 6.12 m |
| 20.00 | 2.7 | -13.0 | -25.3 | -35.1 | -45.6 | 7.50 m | 4.33 m | 3.06 m |
| 30.00 | 7.0 | -7.0 | -18.4 | -28.1 | -38.5 | 5.00 m | 2.89 m | 2.04 m |
| 40.00 | 10.3 | -2.4 | -13.5 | -23.1 | -33.5 | 3.75 m | 2.17 m | 1.53 m |
| 50.00 | 13.2 | 1.2 | -9.6 | -19.2 | -29.7 | 3.00 m | 1.73 m | 1.22 m |
| 60.00 | 15.6 | 4.3 | -6.5 | -16.0 | -26.5 | 2.50 m | 1.44 m | 1.02 m |
| 70.00 | 17.8 | 6.9 | -3.8 | -13.4 | -23.8 | 2.14 m | 1.24 m | 87.48 cm |
| 80.00 | 19.8 | 9.1 | -1.5 | -11.0 | -21.5 | 1.88 m | 1.08 m | 76.55 cm |
| 90.00 | 21.6 | 11.1 | 0.6 | -9.0 | -19.4 | 1.67 m | 96.23 cm | 68.04 cm |
| 100.00 | 23.3 | 13.0 | 2.4 | -7.2 | -17.6 | 1.50 m | 86.60 cm | 61.24 cm |
| 200.00 | 34.7 | 24.9 | 14.4 | 4.9 | -5.6 | 75.00 cm | 43.30 cm | 30.62 cm |
| 300.00 | 41.6 | 31.9 | 21.5 | 11.9 | 1.5 | 50.00 cm | 28.87 cm | 20.41 cm |
| 400.00 | 46.5 | 36.9 | 26.5 | 16.9 | 6.5 | 37.50 cm | 21.65 cm | 15.31 cm |
| 500.00 | 50.4 | 40.8 | 30.3 | 20.8 | 10.3 | 30.00 cm | 17.32 cm | 12.25 cm |
| 600.00 | 53.5 | 44.0 | 33.5 | 24.0 | 13.5 | 25.00 cm | 14.43 cm | 10.21 cm |
| 700.00 | 56.2 | 46.6 | 36.2 | 26.6 | 16.2 | 21.43 cm | 12.37 cm | 8.75 cm |
| 800.00 | 58.5 | 49.0 | 38.5 | 29.0 | 18.5 | 18.75 cm | 10.83 cm | 7.65 cm |
| 900.00 | 60.6 | 51.0 | 40.6 | 31.0 | 20.6 | 16.67 cm | 9.62 cm | 6.80 cm |
| 1000.00 | 62.4 | 52.8 | 42.4 | 32.8 | 22.4 | 15.00 cm | 8.66 cm | 6.12 cm |
| 2000.00 | 74.4 | 64.9 | 54.4 | 44.9 | 34.4 | 7.50 cm | 4.33 cm | 3.06 cm |
| 3000.00 | 81.5 | 71.9 | 61.5 | 51.9 | 41.5 | 5.00 cm | 2.89 cm | 2.04 cm |
| 4000.00 | 86.5 | 76.9 | 66.5 | 56.9 | 46.5 | 3.75 cm | 2.17 cm | 1.53 cm |
| 5000.00 | 90.3 | 80.8 | 70.3 | 60.8 | 50.3 | 3.00 cm | 1.73 cm | 1.22 cm |
| 6000.00 | 93.5 | 84.0 | 73.5 | 64.0 | 53.5 | 2.50 cm | 1.44 cm | 1.02 cm |
| 7000.00 | 96.2 | 86.6 | 76.2 | 66.6 | 56.2 | 2.14 cm | 1.24 cm | 0.87 cm |
| 8000.00 | 98.5 | 89.0 | 78.5 | 69.0 | 58.5 | 1.88 cm | 1.08 cm | 0.77 cm |
| 9000.00 | 100.6 | 91.0 | 80.6 | 71.0 | 60.6 | 1.67 cm | 0.96 cm | 0.68 cm |
| 10000.00 | 102.4 | 92.8 | 82.4 | 72.8 | 62.4 | 1.50 cm | 0.87 cm | 0.61 cm |

Table 3.8-1: RADIATION FROM A 1 $cm^2$ LOOP CARRYING 1 mA OF CURRENT

## 3.9 Conducted Emissions

The purpose here is to quantify conducted noise at the power entry point of a PCB. As discussed in the preceding Section, onboard-generated conducted noise can, via the backplane or motherboard, be impressed on an I/O port and/or the power cord, leading to conducted and radiated emissions difficulties. Typically, the radiation problem so created is much more serious than that caused by direct radiation from power bus traces.

### Conducted Noise at Low Frequencies

By low frequencies is meant the range of frequencies previously defined in (3.3-2):

$$f < \frac{f_d}{[1 + L_b/4ESL_d]^{1/2}} \qquad (f < 11.3\ \text{MHz}) \tag{3.9-1}$$

Then, the worst-case low-frequency model applies (Figure 3.6-1). The practical conditions

$$C_b >> KC_d \qquad (4.7\ \text{uF} >> 0.042\ \text{uF}) \tag{3.9-2}$$

$$C_d >> N(C_s + C_L) \qquad (0.007\ \text{uF} >> 150\ \text{pF})$$

are assumed. Straightforward solution of the frequency-domain loop equations for the model in Figure 3.6-1 leads to the following expression for the ratio of the voltage at the power entry point on the PCB, $V_b(f)$, to that of the square wave (noise) generator, $V_u(f)$ [= Fourier transform of $v_u(t)$]:

$$\frac{V_b(f)}{V_u(f)} \cong \frac{KN(C_s + C_L)}{C_b} \times \left[ \frac{1 - (f/f_b)^2 + j(f/f_{b1})}{1 - (f/f_{cr})^2 + j(f/f_{cr1})} \right], \tag{3.9-3}$$

where

$$f_b = \text{Self-resonant frequency of power entry capacitor} = \frac{1}{2\pi(ESL_bC_b)^{1/2}} \qquad (= 0.733\ \text{MHz}),$$

$$f_{cr} = \text{Resonant frequency of charging loop} \cong \frac{1}{2\pi[\{ESL_b + L_{bbe}(f)\}KC_d(f)]^{1/2}}, \text{ and}$$

$$C_d(f) = \text{Frequency-dependent decoupling capacitance} = C_d/[1 - (f/f_d)^2]^{1/2} .$$

Break frequencies in (3.9-3) due to resistances are

$$f_{b1} = \frac{1}{2\pi ESR_b C_b} \quad (= 169\ \text{kHz}), \qquad f_{cr1} = \frac{1}{2\pi\{ESR_b + R_{fb}(f)\}KC_d(f)} \tag{3.9-4}$$

The above results will be interpreted in conjunction with those developed below for high frequencies.

**Conducted Noise at High Frequencies**

"High frequencies" for the example are those above 11.3 MHz.

Segment-to-segment isolation of the power bus occurring at high frequencies implies that noise at the power-entry point is primarily due to the first DIP in the chain. High-frequency noise due to DIPs further from the power entry point is attenuated by the LC lowpass filter inherent in trace inductances and decoupling capacitors comprising the power bus.

Using [9, pp. 182 - 191], it can be shown that the contribution to noise at the power entry point due to the kth DIP in the chain is proportional to

$$e^{-|a|(k-1)}, \tag{3.9-5}$$

where "a" is the exponent determined in (3.3-3). As discussed in Section 3.3, $|a|$ is large when a good layout and good decoupling capacitors are used. For the example in Table 3.5-2, the worst-case (f --> infinity) value of $|a|$ determined from (3.3-3) is 0.72 and, e.g., high-frequency noise at the power entry point due to the second DIP in the chain is $e^{-0.72} = 50.0$ % of that due to the first DIP in the chain. Noise due to the third DIP in the chain is $e^{-2 \times 0.72} = 23.7$ % of that due to the first DIP, and so on. As the frequency is reduced towards 11.3 MHz, this percentage decreases to a smaller value; at the self-resonant frequency of the decoupling capacitor, all noise at the power-entry point is due to the first DIP in the chain (i.e., the second and subsequent DIPs contribute zero percent to noise).

The model for analysis of high-frequency conducted emissions is illustrated in Figure 3.9-1. That model is derived from the beginning segment model of Figure 3.3-1A with the paralleling of inductors indicated in Figure 3.3-2. Also, the model of the DIP given in 3.2-2B is added. The model in Figure 3.9-1 neglects contributions to high-frequency noise from DIPs other than the first DIP in the chain. Although useful for the present purposes, this model provides an increasingly optimistic estimate of noise as the frequency is increased above the self-resonant frequency of the decoupling capacitor.

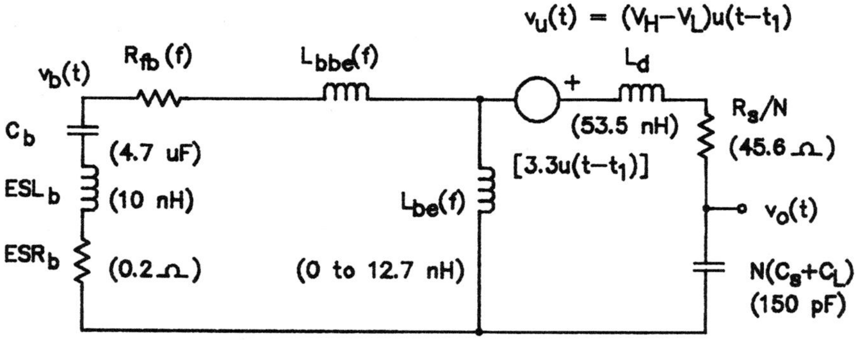


Figure 3.9-1: MODEL FOR HIGH-FREQUENCY CONDUCTED EMISSIONS

Straightforward analysis of the loop equations for the model leads to the following expression for the ratio of noise at the power entry point to that generated by the square-wave switching signal, $v_u(t)$:

$$\frac{V_b(f)}{V_u(f)} \cong \frac{j2\pi f ESL_b}{ESR_b + R_{fb}(f)} \times \frac{(j2\pi f)^2 L_{be}(f) N (C_S + C_L)}{(1 + jf/f_{d2})[1 - (f/f_{dr})^2 + jf/f_{dr1}]}, \qquad (3.9\text{-}6)$$

where, in addition to those of (3.9-2), the approximation $2\pi f ESL_b >> ESR_b$ has been used (at high frequencies, this is nearly always true in practice). In this expression,

$$f_{dr} = \text{Resonant frequency of discharge loop} = \frac{1}{2\pi\{L_d N(C_S + C_L)\}^{1/2} A(f)} \qquad (\cong 56.1 \text{ MHz}),$$

$$A(f) = \left[1 + \frac{L_{be}(f)/L_d}{1 + L_{be}(f)/(L_{bbe}(f) + ESL_b)}\right]^{1/2} \qquad (\cong 1),$$

$$f_{dr1} = \frac{1}{2\pi}\left[\frac{1}{[R_S(C_S + C_L)}\right] \qquad (\cong 23.2 \text{ MHz}),$$

$$f_{d2} = \frac{1}{2\pi}\left[\frac{R_{fb}(f) + ESR_b}{L_{bbe}(f) + ESL_b + L_{be}(f)}\right].$$

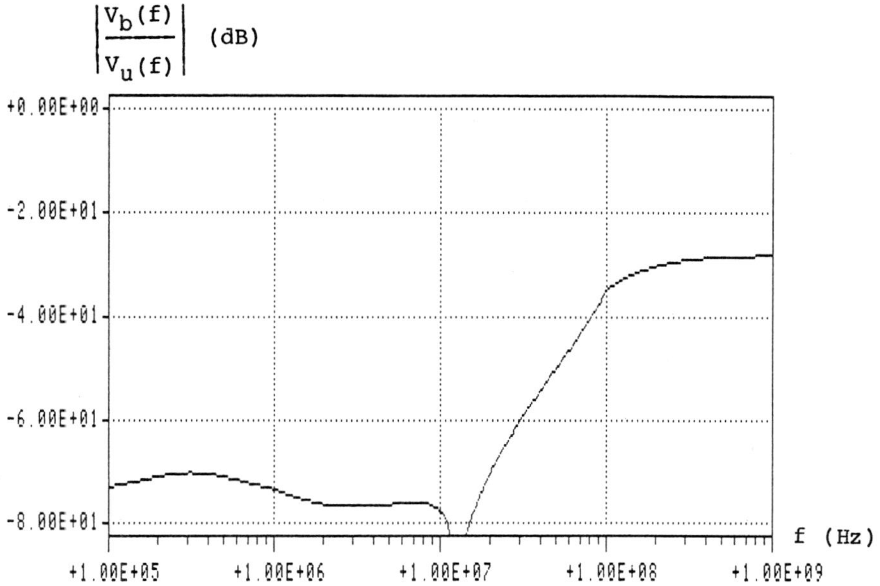


Figure 3.9-2: CONDUCTED NOISE

For $f >> f_{dr}$, the ferrite bead no longer provides filtering and the attenuation of the noise is purely due to inductances; (3.9-3) reduces to

$$\frac{V_b(f)}{V_u(f)} \simeq \frac{L_{be}(f)\ ESL_b}{L_d(L_{bbe}(f) + L_{be}(f) + ESL_b) + (L_{bbe}(f) + ESL_b)L_{be}(f)} \quad (= -\ 28.3\ \text{dB}), \tag{3.9-7}$$

From (3.3-5), $L_{be}(f)$ is proportional to $ESL_d$, the sum of the internal self-inductance of the decoupling capacitor and the inductance of the leads used to connect that capacitor to the power bus. Thus, it is clear from (3.9-7) that, as the frequency is increased beyond the resonant frequency of the discharge loop, noise suppression is dependent on the inductance, and not the capacitance, of the power entry and decoupling capacitors.

**Interpretation**

Initial interpretation will be in terms of numerical results given in Figure 3.9-2 for the example. Subsequent interpretation will illustrate the effects of deviations from the (good) design principles employed in that example.

The condition

$$\frac{|V_b(f)|}{|V_u(f)|} \ll 1 \qquad (3.9\text{-}8)$$

is desirable at all frequencies, as transfer of noise from the PCB to the power entry point is small when this condition prevails.

For the purposes of computation of both low and high frequency results for Figure 3.9-2, the characteristics of the ferrite bead cited in Figure 3.2-5 and Appendix 3 have been approximated as a series of straight-line segments on a logarithmic frequency scale:

(3.9-9)

$$L_{fb}(f) = \begin{cases} 3.18 - 4 \log (f/10^6) \text{ uH}, & 10^5 < f < 4 \times 10^6 \text{ Hz}, \\ 0.80 - 0.572 \log [f/(4 \times 10^6)] \text{ uH}, & 4 \times 10^6 < f < 10^8 \text{ Hz}, \\ 0, & f > 10^8 \text{ Hz}, \end{cases}$$

$$R_{fb}(f) = \begin{cases} 20 + 20 \log (f/10^6) \text{ Ohms}, & 10^5 < f < 10^6 \text{ Hz}, \\ 20 + 40 \log (f/10^6) \text{ Ohms}, & 10^6 < f < 10^7 \text{ Hz}, \\ 60 - 20 \log (f/10^7) \text{ Ohms}, & 10^7 < f < 10^8 \text{ Hz}, \\ 40 \text{ Ohms}, & f > 10^8 \text{ Hz}. \end{cases}$$

Low Frequencies

As modeled here, noise output from a PCB to the backplane or motherboard occurs at clock and clock-harmonic frequencies. Assuming that the model is "complete," the only noise suppression of interest in Figure 3.9-2 is that corresponding to harmonics of the 4 MHz clock. However, clock frequencies lower than that employed for the example are used. Also, there oftentimes occurs on a PCB noise sources at frequencies considerably lower than the clock frequency (e.g., baud rate generation). For these reasons, discussions following encompass frequencies which are lower than the clock frequency used in the example.

Most power supplies are well-regulated at least into the hundreds of Hertz frequency range. In good designs, the impedance of the backplane/motherboard does not impugn low-frequency regulation at the PCB loads. Then, at those extremely low frequencies, the power supply and outboard DC bus impedances control noise amplitude at the power entry point; that lower impedance is across the power entry capacitor. This is not evident in the present analysis due to the assumption of infinite backplane/motherboard impedance, which is appropriate for worst-case analyses at higher frequencies.

At some frequency where the power supply loses regulation - generally in the 1 kHz to 100 kHz range - the value of the power entry capacitor becomes important for maintaining low noise output from the PCB. If the PCB - or other PCBs/subsystems supplied by the same backplane/motherboard power bus to which the PCB is connected - do not generate noise at those low frequencies, noise suppression is not a problem, and, except for the issue cited below, the value of $C_b$ can be very small. However, an assumption in derivation of the damping equations given in Section 3.6 is that the value of the power entry capacitor is at least ten times the sum of all the decoupling capacitors in a chain of DIPs:

$$C_b \geq 10 \times KC_d \, . \tag{3.9-10}$$

Reduction below this minimum value for the power entry capacitor can effectively reduce the value of the individual decoupling capacitors, thereby causing charging loop ringing. The 9.1 % reduction - computed by determining the capacitance of $C_d$ in series with 10 $C_d$ - implied by the factor of 10 in (3.9-10) is well within the tolerance/conservatism of good designs. For (increased $ESL_b$) reasons discussed below, it can be unwise to select the value of the power-entry capacitor to be, say, another order of magnitude larger than that recommended in (3.9-10).

At frequencies below the self-resonant frequency of the power entry capacitor, $f_b$ (= 0.733 MHz), suppression of noise is largely determined by $ESR_b$, $C_b$, and the ferrite bead. Above that frequency, but below the self-resonant frequency of the decoupling capacitor, (3.9-3) approaches

$$\frac{V_b(f)}{V_u(f)} \cong \frac{(j2\pi f ESL_b + ESR_b)(j2\pi f [C_s + C_L])}{1 - (f/f_{cr})^2 + j(f/f_{crl})} \, . \tag{3.9-11}$$

The numerator term due to the inductance of the power-entry capacitor and connecting leads, $ESL_b$, is greater than $ESR_b$ at relatively low frequencies when a low-$ESR_b$ (i.e., good) power-entry capacitor has been selected. For the example, $ESL_b$ predominates at frequencies above 3.18 MHz.

Once the value of $C_b$ is selected using (3.9-10) and the equivalent series resistance, $ESR_b$, is held to below about one Ohm, the most important consideration is $ESL_b$. A frequent design error is to select a large-valued (and commensurately expensive) power entry capacitor in order to improve noise suppression. From (3.9-11), **the inductance, $ESL_b$, and resistance, $ESR_b$, and not the capacitance, $C_b$, determines noise suppression provided by power entry capacitors at clock and clock-harmonic frequencies of primary interest in modern microprocessor-based equipment.** Oftentimes, a smaller-valued power-entry capacitor will perform better than a larger value due to smaller self-inductance, $ESL_b$, associated with a smaller physical size.

For the results shown in Figure 3.9-2, component values have been selected so that all frequencies below the self-resonant frequency of the decoupling capacitor are attenuated at least 60 dB; this is considered adequate for even the most demanding designs. An approximate upper bound estimate of the maximum 4 MHz (clock fundamental) ripple corresponding to that suppression is 5 volts/1000 = 5 mV peak (10 mV pp).

High Frequencies

As indicated in Figure 3.9-2, noise suppression is large at the self-resonant frequency of the decoupling capacitor, $f_d$ (= 12.3 MHz). At that frequency, emissions will be zero when an infinite-Q (zero-$ESR_d$) decoupling capacitor is used, as the impedance of the capacitor at that frequency is zero.

The effects of nonzero equivalent series resistance of the decoupling capacitor, $ESR_d$, are shown in Figure 3.9-3. Degradations in suppression of noise near $f_d$ due to nonzero $ESR_d$ have been determined from an analysis of the low-frequency equivalent circuit (Figure 3.6-1) with $ESR_d$ substituted for the (resonant) decoupling capacitor. $ESR_d < 0.5$ Ohms is a design recommendation in Section 4.2.

Ringing of the discharge loop causes a peak in the noise transfer function; this does not occur in Figure 3.9-2 because care has been taken to insure that the number of ongoing gates serviced by the decoupling capacitor (N = 5) is low enough to insure overdamping of the discharge loop. An increase in conducted emissions near the ringing frequency (≅ 32 MHz) is apparent for the N = 15 curve of Figure 3.9-4. From (3.7-2), N = 5 corresponds to an overdamped condition (no high-frequency ringing), and N = 15 corresponds to (undesirable) ringing. Radiated emissions caused by 32 MHz noise on the power bus can be damaging, as the "antennas" associated with typical digital circuitry begin to become efficient at that frequency (see Section 3.8).

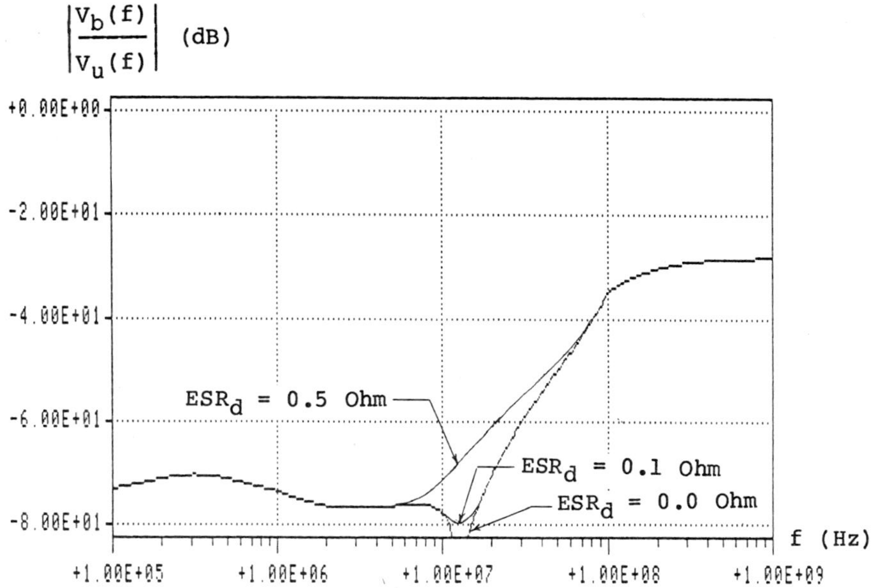


Figure 3.9-3: EFFECT OF EQUIVALENT SERIES RESISTANCE OF DECOUPLING CAPACITOR

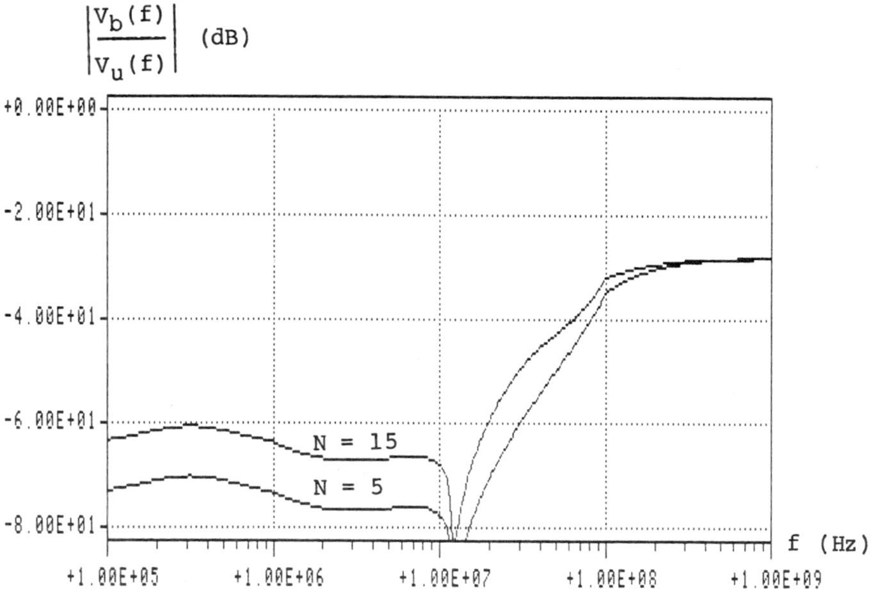


Figure 3.9-4: EFFECT OF TRYING TO SERVICE TOO MANY GATES WITH ONE DECOUPLING CAPACITOR (DISCHARGE LOOP UNDERDAMPING @ N = 15) [$ESR_d$ = 0]

The leading term in (3.9-6) indicates that noise suppression is enhanced by the resistance of the ferrite bead, $R_{fb}(f)$, with this effect aided by a low-$ESL_b$ power entry capacitor. (3.9-7) indicates that **suppression of very high frequency noise is aided by the inductances $L_{bb}$, $L_b$ and $L_d$.**

In the foregoing Figures, the increase in conducted noise starting at the self-resonant frequency of the decoupling capacitor is primarily due to the increase in impedance of $ESL_b$ and $ESL_d$ with increasing frequency. Reduction of either $ESL_d$ or $ESL_b$ results in a commensurate decrease in conducted noise at frequencies above the self-resonant frequency of the decoupling capacitor.

Except in second-order ways, **high-frequency conducted emissions suppression does not depend on the values of the decoupling or power entry capacitors, but, rather, on the equivalent series inductance (ESL) associated with those capacitors.** Smaller $ESL_b$ and $ESL_d$ reduce high-frequency noise. As defined here, those inductances also include the leads and traces necessary to connect those capacitors across the power bus; clearly, short leads and wide traces are in order.

**Results for Noise Suppression Without a Ferrite Bead**

Figure 3.9-5 corresponds to the example as stated in Table 3.5-2 except that the ferrite bead is removed. All other elements are the same. As is easily verified from (3.6-3), the charging loop is underdamped ($D_b$ = 89.1 > 1 @ f = 4 MHz). Then, the circuit will ring at approximately 4 MHz, corresponding to the low-frequency peak in the noise transfer curve shown in Figure 3.9-5.

An upper bound on the amplitude of the ripple is computed as follows: 30 dB below 5 volts = 5/31.6 = 158 mV peak (= 316 mV pp) ripple. In addition to causing excessive conducted emissions at the fundamental clock frequency, 4 MHz (which, depending on power supply filters, could cause out-of-specification emissions at the power cord), the (158 mV) peak amplitude of the ripple begins to approach the (300 mV) noise immunity limit for LSTTL given in Table 3.4-1; the gates are more susceptible to "soft" logic errors.

Figure 3.9-6 illustrates the results of an attempt to overcome the lack of a ferrite bead by using a larger-value decoupling capacitor; $C_d$ has been increased from 0.007 uF to 0.20 uF, while maintaining the same $ESL_d$ (=24.8 nH). Relative to the "Without Ferrite Bead" curve of Figure 3.9-5, this has substantially improved low-frequency conducted noise and eliminated the susceptibility problem. However, the decoupling capacitor is more expensive and, more importantly, the frequency of maximum noise attenuation - the self-resonant frequency of the decoupling capacitor - has been reduced to a

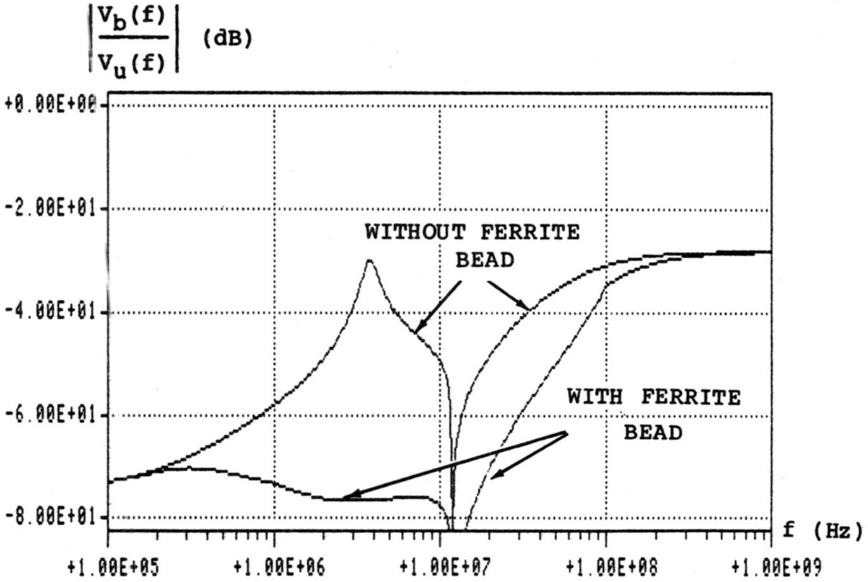


Figure 3.9-5: EXAMPLE CIRCUIT WITH AND WITHOUT A FERRITE BEAD ($C_d$ = 0.007 uF)

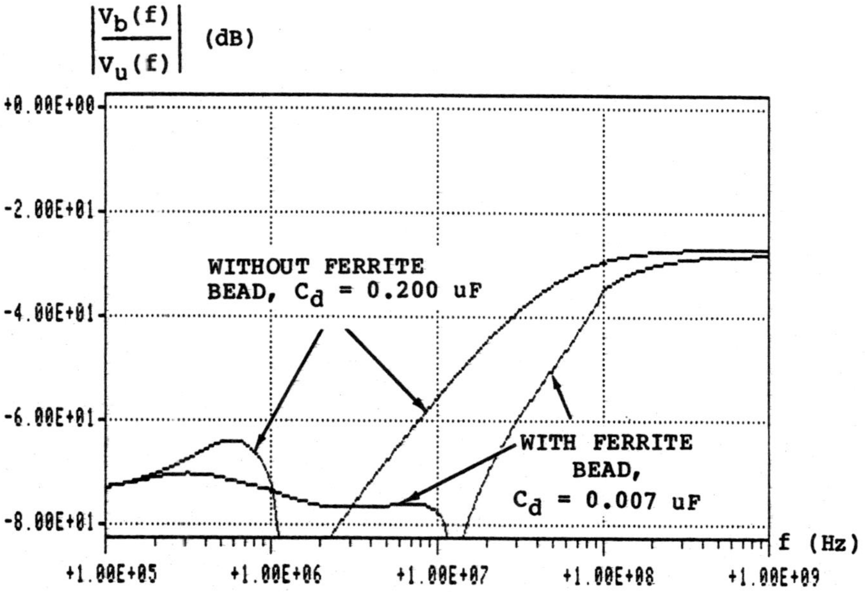


Figure 3.9-6: NO FERRITE BEAD, BUT LARGER-VALUED DECOUPLING CAPACITOR

frequency where noise suppression at dangerously emissive frequencies in the range of a few tens of MHz to 300 MHz is, relative to the "With Ferrite Bead" curve of Figure 3.9-5, greatly reduced.

One way to suppress troublesome radiated emissions caused by noise on the power bus is to select the decoupling capacitor so that resonance, including the effects of leads/traces, occurs at or near troublesome frequencies. Although the minimum lower bound on $C_d$ (3.5-3) should be observed in the process, **a smaller-valued decoupling capacitor will, due to a higher self-resonant frequency, oftentimes perform better than a larger (and perhaps more expensive) value.** Usually, damping considerations cited in (3.6-4) permit this "tuning" approach if and only if a ferrite bead is used.

**In general, the self-resonant frequency of decoupling capacitors should be as high as damping considerations, PCB layout, and capacitor technology permit.** This places the resonant null in the ten to 100 MHz range where emissions are most troublesome.

It is recommended in Section 4.2 that **a ferrite bead should be used in all designs.**

**General Results and Comments On Low-Impedance Power Bus Products**

An introduction to low-impedance power bus products is best made by stating general results for high frequencies apparent from the foregoing analysis.

**$ESL_d$ should be minimized.** Methods for doing this include specification of low-inductance capacitors (and the lead lengths used to measure that inductance) and laying out the PCB so as to provide for minimum lead and trace lengths for connecting the those capacitors across the power bus.

It is clear from (3.9-7) that the PCB inductances $L_{bb}$, $L_b$, and $L_d$ help filter high-frequency noise, as increases in those inductances reduce $|V_b(f)|/|V_u(f)|$. However, it is equally clear from (3.7-2) that increased $L_d$ decreases the number of gates servicable by a decoupling capacitor without causing high-frequency ringing.

Then, in contrast to the frequently-recommended approach of maintaining the impedance (inductance) of the power bus as small as possible, **emissions considerations suggest that modest to moderate $L_{bb}$ and $L_b$ components of the power bus inductance are desirable.** A ferrite bead is used to (easily) prevent charging loop ringing problems which could otherwise occur due to modest $L_{bb}$ and $L_b$.

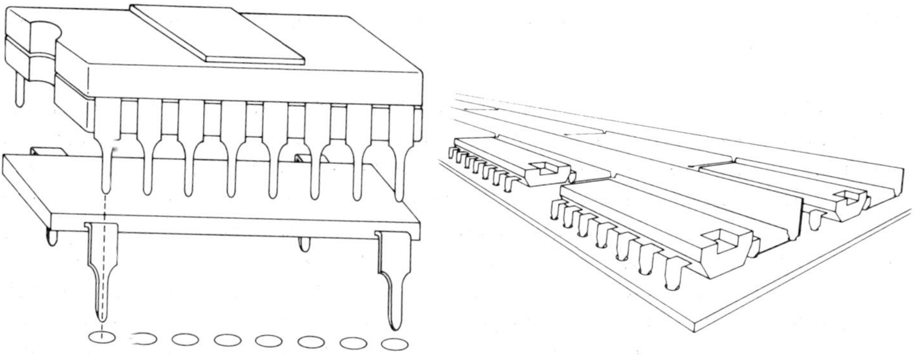

Under-The-DIP                    Strip-Type Bus

Figure 3.9-7: LOW-IMPEDANCE POWER BUS PRODUCTS (Courtesy Rogers Corporation)

Commercially-available under-the-DIP (and in-the-DIP-socket) decoupling capacitor products usually reduce the inductance of decoupling capacitors to a level not achievable by conventional means. However, it is the total (trace/lead plus intrinisic capacitor) inductance incurred in connecting the decoupling capacitor across the power bus, and **not** across the DIP, (i.e., $ESL_d$, **not** $L_d$, which is more important to minimize. PCB layout to minimize inductance incurred in connecting a decoupling capacitor across the power bus can be more important - and less expensive - than the low intrinsic inductance capacitors embodied in the products mentioned above.

By minimizing $L_d$, under-the-DIP and strip-type products permit those products to service a greater number of output gates without ringing than is usually achievable via conventional means. However, by decreasing $L_d$, the effectiveness of filtering high-frequency noise by the power bus is also decreased. Also, there is a minimum lower bound on $L_d$ set by the inductance of the wire inside the DIP which connects the output gates on the chip to the pins of the DIP. If a PCB is carefully laid out, conventional traces do not significantly increase the $L_d$ beyond that intrinsic to the DIP package.

The desirability of modest to moderate $L_{bb}$ and $L_b$ conflicts with the concept of a number of commercial "low-Z" strip-type distributed power bus products. Those products can reduce $ESL_d$ to a value which is difficult to achieve with discrete components. However, a part of the power bus inductance which helps filter noise, $L_{bb}$ and $L_b$, is also reduced. For this, cost, and fragility (manufacturing reliability, particularly during wave soldering) reasons, strip-type products should be carefully scrutinized. It is not clear that, relative to a conventional layout with good components, reduction of power bus noise **at the power entry point** is achieved by those products, and they are, relative to discrete decoupling capacitors and PCB traces, expensive.

However, outside of decoupling considerations, strip-type products can reduce ground noise (ground inductance) to a level which is not achievable via conventional PCB layout techniques: in attempting to achieve the small ground inductance of a strip-type bus product, the wide traces which would have to be used in a conventional layout could consume substantial PCB area, thereby reducing the component density below that which may be achieved with strip-type power bus products. Strip-type products which intersect the PCB at right angles provide, in effect, another physical dimension of the PCB for power and ground bus use and, thus, greater component density. However, metallic strips at right angles to the PCB also achieve the goals of providing a third dimension for ground bus distribution and a small ground inductance (and noise), and are used by at least one large manufacturer of microcomputers, apparently for that purpose.

From a radiated emissions standpoint, both under-the-DIP and strip-type power bus products have the desirable effect of maintaining the "loop antenna" area of the power bus at a very low value. However, loop-based radiation from the power bus is not usually the worst radiation problem associated with inadequate decoupling - clock loops tend to dominate loop-based radiation. Rather, those radiated emissions difficulties discussed in Section 3.8 which stem from conducted power bus noise are more troublesome.

In minimizing radiated emissions difficulties, it is more important to decrease conducted power bus noise and direct radiation from clock loops than it is to minimize radiation from the power bus.

Careful layout, a ferrite bead, and an adequate number of high-quality decoupling capacitors will oftentimes achieve the goals of the above special products at significantly reduced cost.

## 3.10 Variations with Logic Families

The purpose here is to indicate the relative difficulty of decoupling ECL, CMOS, HCMOS, and LSTTL logic families.

Preceding analyses are independent of the logic family; in particular, definitions of low and high frequencies do not depend on the logic family {see (3.2-1) and (3.2-2)}. From those analyses, it is apparent that there are two damping concerns: If the charging loop is underdamped, low-frequency ringing (usually below 30 MHz) occurs. If the discharge loop is underdamped, high-frequency ringing (usually above 30 MHz) occurs.

The undesirable pulse integrity/susceptibility and emissions consequences of underdamping (ringing) have been discussed in preceding Sections.

**Charging Loop Damping**

The purpose here is to exemplify variations in charging loop damping due to different logic families. For that purpose, the design recommendation $C_d = 9N[C_S + C_L]$ is used {see (3.5-3)}. It is also assumed that, as is true for most designs where the ferrite bead has been properly selected, $L_{bbe}(f) \approx L_{fb}(f)$ and $ESR_b << R_{fb}(f)$ for frequencies where ringing would otherwise be present. Then, (3.6-3) reduces to

$$D_b \cong \frac{0.4\, L_{fb}(f)}{R_{fb}^2(f) KN[C_S + C_L])} < 1, \qquad (3.10\text{-}1)$$

from which it is apparent that, for a given layout and ferrite bead, differences in dampings for various logic families are due to differences in the sums of the gate output and input capacitances. Those capacitances are given in Table 3.2-1:*

LSTTL: $C_S + C_L = 25$ pF, CMOS: $C_S + C_L = 15$ pF, HCMOS: $C_S + C_L = 34$ pF.

Overdamping of conventional relatively slow-speed CMOS is the most difficult to achieve. However, the approximately 2:1 variation in $C_S + C_L$ for

---

* The expression (3.10-1) does not apply to ECL, as the loads, at frequencies where charging loop ringing is likely to occur, are primarily resistive (50 Ohms), not capacitive. By using the model of Figure 3.6-1 with a resistance of 50/KN Ohms across the capacitance $KN(C_S + C_L)$, it is straightforward to demonstrate that charging loop ringing is less likely to occur than it is for LSTTL, CMOS, and HCMOS logic families.

logic families is easily overcome by selection of a proper ferrite bead. In practice, the bead can be chosen so that $R_{fb}(f)$ leads to overdamping ($D_b$ < 1) regardless of the logic family.

**Discharge Loop Damping**

From (3.6-2), the maximum number of gates servicable by a single decoupling capacitor without incurring discharge-loop ringing is

$$N < \frac{R_s^2(C_s + C_L)}{4[L_{be}(f) + L_d]} \qquad (3.10\text{-}2)$$

There is a strong dependence of damping of the discharge loop on whether ECL, CMOS, LSTTL, or HCMOS is used. That dependence is not mitigated by a ferrite bead.

Example: The numerator in the expression above is dependent on the logic family ($R_s$, $C_s$, and $C_L$). Those data are available from Table 3.2-1. The denominator does not depend on the logic family, but, rather, on the layout. The values $L_d$ = 53.5 nH and $L_{be}(\infty)$ = 12.7 nH applicable to the layout in Figure 3.2-1 are used. Then, substitution into (3.10-2) leads to

LSTTL: $N \leq 6$, CMOS: $N \leq 17$, HCMOS: "N" $\leq$ 0.219, ECL: "N" $\leq$ 0.025

At least for the typical (good-but-not-excellent design) inductances used, it is clear that HCMOS and ECL logic families will ring even for one output gate. Those families rely on pullup/pulldown resistors to prevent ringing of the charging loop. Those resistors are not taken into account by the model.

For example, the conventional termination for ECL output lines, 50 Ohms, appears directly across the gate output capacitance, $C_s$ (= 13 pF). Then, each gate load is primarily resistive up to

$$f = \frac{1}{2\pi RC_s} = \frac{1}{2\pi(50)(13 \times 10^{-12})} = 244.6 \text{ MHz.} \qquad (3.10\text{-}3)$$

Due to the terminating resistor, there is no discharge loop damping problem over the useful frequency range of the gate.

From the standpoint of not requiring pullup/pulldown resistors in order to prevent ringing in the discharge loop, LSTTL is to be preferred over HCMOS. There are, of course, other considerations - such as speed and power consumption - governing selection of a logic family. The point to be made is that the power dissipation of pullup/pulldown resistors and increased component count consequences of selecting a logic family requiring those resistors should be taken into account in selecting a logic family.

### 3.11 Sharing a Single Decoupling Capacitor Amongst Multiple DIPs

Answers to the question "How many DIPs can share one decoupling capacitor without incurring undesirable effects?" are important to economics and PCB layout. Fewer decoupling capacitors implies more PCB area for other purposes and a lower total cost for decoupling capacitors. The purpose here is to explore limitations of representative sharing arrangements.

Figure 3.11-1 illustrates several means by which a single decoupling capacitor can be shared between two or more DIPs.

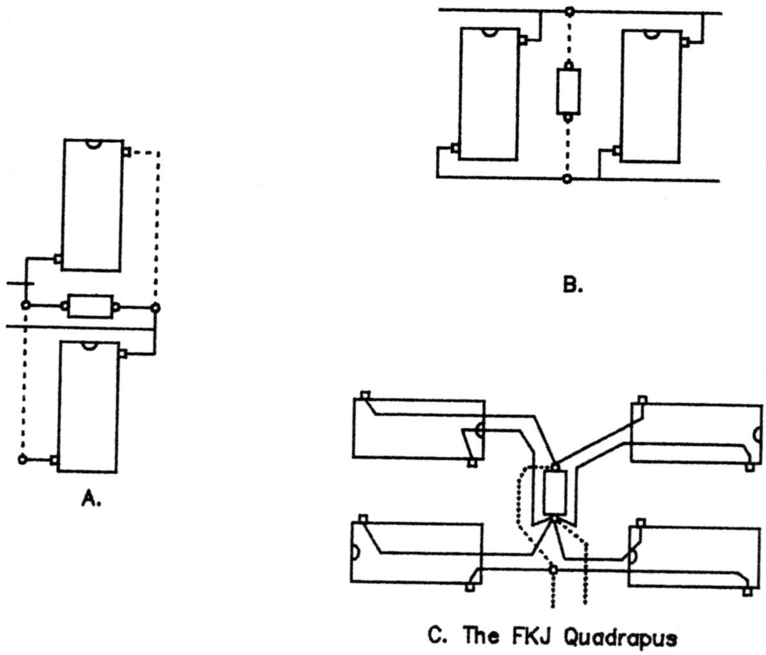


―――――――― = Component Side      ......... = Solder Side

Figure 3.11-1: WORST-CASE (END-SEGMENT) SHARING OF A DECOUPLING CAPACITOR

For all cases in the Figure, the equivalent circuit of the power bus is unchanged. Only the equivalent circuit of what is connected across the power bus is changed by the addition of DIPs. In Figure 3.11-1 A and B, $ESL_d$ and $L_d$ (per DIP) is not changed from the one-DIP case by the added DIP.

**Sharing at Low Frequencies**

Let

P = Number of Decoupling Capacitors supplying K DIPs $\leq$ K (3.11-1)

The requirement (3.6-3) for damping of the charging loop becomes

$$D_b = \frac{4(L_{bbe}(f) + ESL_b)}{(R_{fb}(f) + ESR_b)^2 PC_d} < 1, \tag{3.11-2}$$

where

P = Number of decoupling capacitors $\leq$ K = Number of DIPs

For P < K, $D_b$ is, relative to P = K, increased; the circuit tends towards ringing. However, the decrease in the capacitance term in the denominator can be overcome by using larger-valued decoupling capacitors. Additionally, the ferrite bead can, within broad limits, be selected to increase damping. The practical upper limit of the value of the decoupling capacitors is set by bounds on the physical size of those capacitors and, usually, the increase in $ESL_d$ associated with larger-sized capacitors.

**Sharing at High Frequencies**

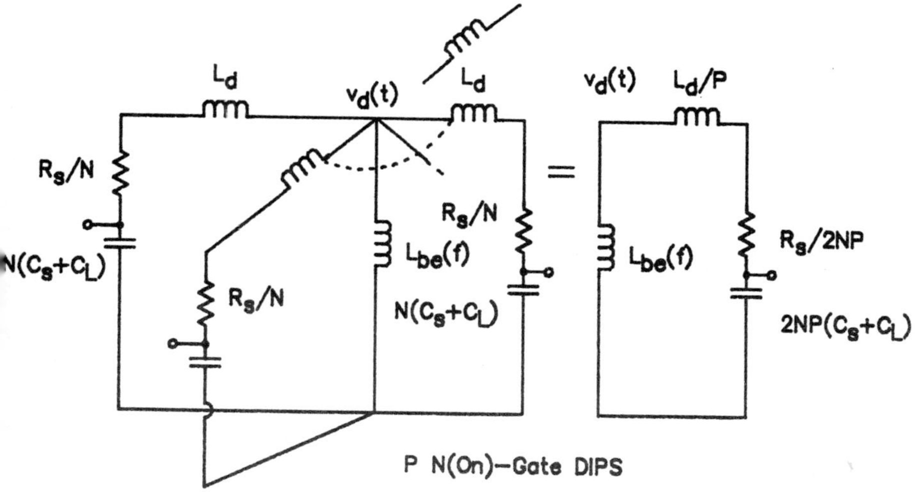


Figure 3.11-2: HF EQUIVALENT CIRCUIT FOR LAYOUTS OF FIGURE 3.11-1

Figure 3.11-2 is the high-frequency equivalent circuit for sharing arrangements involving the worst-case end-segment DIP. This Figure is derived from Figure 3.3-2B, with $L_{be}(f)$ as given in (3.3-5). P DIPs are sharing a single decoupling capacitor, and each of those DIPs have N gates going on at once. The inductance incurred in connecting each DIP across the decoupling capacitor is assumed to be equal to $L_d$.

By applying the expression for the underdamped condition given in Table 3.5-2 to the circuit shown in Figure 3.11-2, it is easily shown that the number of DIPs, each with N output gates going on, which can share a single decoupling capacitor without incurring high-frequency ringing is

$$P < \frac{R_s^2(C_s + C_L)}{4N\,L_{be}(f)} - \frac{L_d}{L_{be}(f)} \qquad (3.11\text{-}3)$$

In those cases where the right-hand side of this expression is greater than 2, an optimistically large estimate has been provided if $L_d$ is based on the one-DIP-per-capacitor value of $L_d$. For $P > 2$, it is a topological certainty that $L_d$ must increase beyond that required for the cases $P = 1, 2$*; i.e., for $P > 2$, $L_d$ is an increasing function of P. That increase is due to the additional length of DIP-to-capacitor traces necessary to connect $P > 2$ DIPs to the decoupling capacitor. Three-dimensional packaging and layouts would help for the cases $P > 2$, but that packaging is not currently in widespread use.

**Example**

For the example of Table 3.5-2, $L_b = 13.3$ nH, $L_d = 53.5$ nH, $C_s + C_L = 30$ pF, and $ESL_d = 24.8$ nH. From (3.3-5), the maximum ($[f/f_d]^2 >> 1$) value of $L_{be}(f)$ is 12.7 nH. Then, (3.11-3) reduces to

$$P < \frac{30.7}{N} - 4.21 = 1.93 \text{ @ } N = 5. \qquad (3.11\text{-}4)$$

Only one DIP per decoupling capacitor is permitted.** If, however, the number of output gates per DIP is reduced from 10 to 8 (i.e., N = 5 is changed to N = 4), 3 DIPs could share the same decoupling capacitor. How-

* Figures 3.11-1A and B illustrate two P = 2 cases where $L_d$ per DIP is not increased beyond that for the case P = 1.

** Due to the closeness of P to 2, two DIPs would not cause severe problems.

ever, because $L_d$ is likely to increase with addition of the third DIP, 3 is an optimistic value; 2 is more likely when the necessary increase in $L_d$ is taken into account.

**Discussion**

The low-frequency requirement (3.11-2) is not particularly restrictive, as the value of the decoupling capacitor is generally chosen to provide some degree of margin beyond the minimum required value.

The high-frequency requirement (3.11-3) is much more restrictive, and sets the upper limit of P to one or two in most practical cases. To increase P to larger values without incurring VHF ringing and the undesirable emissions consequences of that ringing, it is necessary to exercise extreme care in PCB layout and component selection to minimize both $ESL_d$ and $L_d$.

Even with careful layout, the maximum number of DIPs which can share a single decoupling capacitor without incurring undesirable VHF emissions performance is, in most cases, two. This limitation stems from a relatively large increase in $L_d$ attendant to a two-dimensional grouping of more than two DIPs.

## 3.12 Extensions of Models to Multiple Rows of DIPs

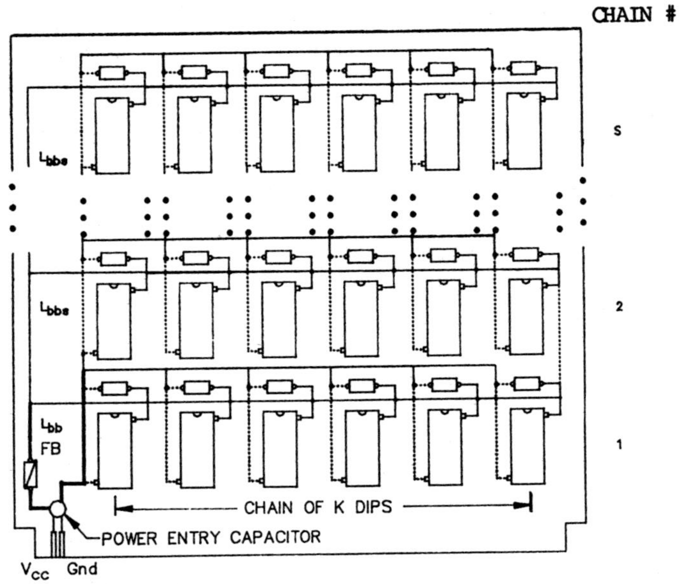


Figure 3.12-1: S ROWS OF K-DIP CHAINS

In this Section, models developed in Section 3.2 will be extended to incorporate multiple K-DIP chains:

$$S = \text{Number of K-DIP chains} = 1, 2, 3, \ldots \qquad (3.12\text{-}1)$$

It is assumed that

- Each chain has K DIPs with each DIP having N gates going on at once,
- There is one DIP per decoupling capacitor in each chain,
- All decoupling capacitors are equal in value and $ESL_d$, and that
- The length of power feed traces between rows of DIPs is the same.

Key inductances are those defined in Figure 3.2-1, equation (3.1-1) and

$L_{bbs}$ = Inductance of power feed traces between rows of K-DIP chains.

**The aim here is to show that worst-case design is for a single (S = 1) chain of DIPs.** The primary effects of additional chains of DIPs are to reduce conducted noise and make overdamping of the overall charging loop easier than for a single chain of DIPs. With modifications for low frequencies which accomodate the larger total capacitance ($C_b \geq 10KC_d \rightarrow C_b \geq 10SKC_d$), results for a single chain of DIPs derived in preceding Sections should be used for worst-case design.

Radiated noise considerations for S > 1 do not change from those for the case S = 1 discussed in Section 3.8. Additional DIPs are being added, but so are additional clock loops. Direct radiation from clock loops is still greater than direct radiation from the power bus for all but the most poorly-designed power buses.

**Conducted Noise From Multiple K-DIP Chains at High Frequencies**

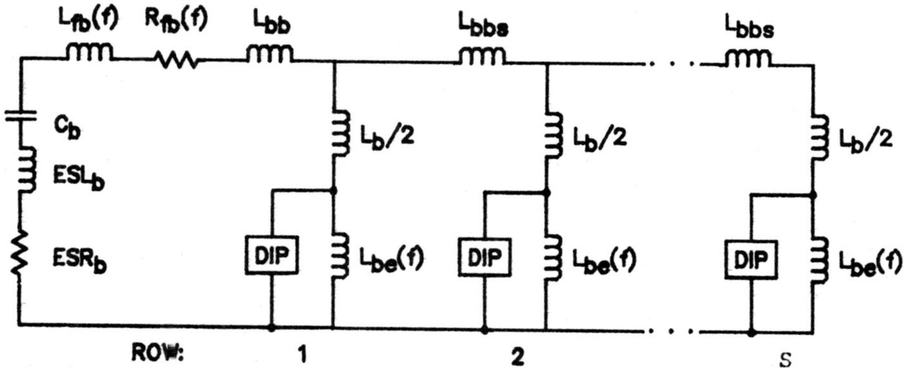


A. Power Bus for S K-DIP Chains

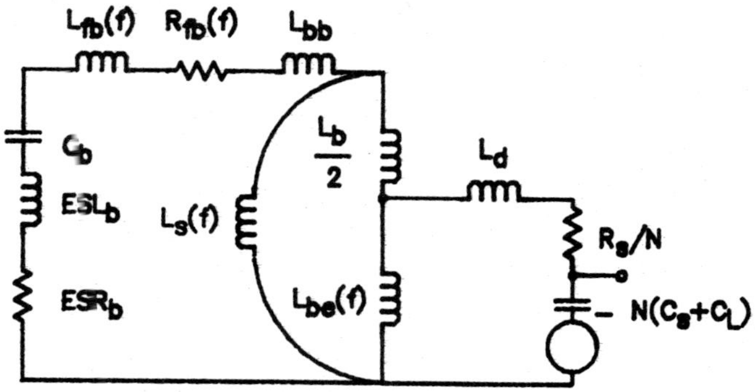


B. First DIP in First Chain of K DIPs and Bus Impedance Seen by That Chain

Figure 3.12-2: HIGH-FREQUENCY MODELS

Figure 3.12-2A is the high-frequency equivalent circuit for the S-chain case. As a condition which prevails in all practical circuits, it is assumed that loading of the power bus by the DIPs is small: $2\pi f L_{be}(f) << DIP$ impedance. Then, the input impedance of each of the K-DIP chains at high frequencies is accurately represented by the combined effective decoupling capacitor and inter-DIP bus inductance, $L_{be}(f)$, with the value of that inductor given in (3.3-5).

Since, in practice, $L_{bbs} >> L_b/2 + L_{be}(f)$, the voltage divider action of inter-DIP power feed traces ($L_{bbs}$) and $L_b/2 + L_{be}(f)$ tends to suppress conducted noise from all but the first chain of DIPs. It was shown in Section 3.9 that, within that first chain, the beginning DIP segment is the dominant contributor to noise.

The model in Figure 3.12-2B pertains to the first DIP in the first K-DIP chain. The inductance $L_s(f)$, to be derived below, represents the impedance of the rest of the K-DIP chains at high frequencies.

By the same methods used to used to develop results for the single chain of K DIPs in Section 3.2, the equation

$$a = \operatorname{arccosh} \left[1 + \frac{L_{bbs}}{L_b + 2L_{be}(f)}\right], \quad \text{real,} \tag{3.12-2}$$

is derived in Appendix 1. The condition

$$e^{S|a|} >> e^{-S|a|} \tag{3.12-3}$$

usually prevails in practical circuits. Then, the impedance of the power bus seen at the juncture of the $L_{bb}$ and $L_{bbs}$ traces, looking towards the 2nd, 3rd .. Sth chain of DIPs, is equivalent to that of the inductor

$$L_s(f) = \frac{L_{bbs}}{2} \left[1 + \left\{1 + \frac{4L_{be}(f) + 2L_b}{L_{bbs}}\right\}^{1/2}\right] \tag{3.12-4}$$

The basis for this result is derived in Appendix 1.

Figure 3.12-2B is similar to Figure 3.9-1 pertinent to the single-chain case with $L_s(f)$ added. Relative to that present in the single-chain case, $L_s(f)$ serves to decrease conducted noise present at the power entry point by lowering the impedance at the juncture of the power-entry power feed traces and the beginning of the inter-DIP power feed traces. In good designs, however, $L_s(f)$ is usually much greater than $L_b/2 + L_{be}(f)$ and reduction in noise from that present in to the single-chain case is minimal. Thus, use of results given in Section 3.9 for estimating high-frequency conducted noise for a single K-DIP chain, while providing a worst case, does not result in extreme overdesign.

Using the same logic as was employed at the beginning of the discussion of high-frequency noise in Section 3.9, it can be shown that, particularly in view of the relatively large value of "a" in (3.12-2) which is obtained in good designs, noise at the power-entry point due to the second and subsequent chains of DIPs is minimal.

**Charging Loop (Low Frequency) Damping for Multiple Rows of K-DIP Chains**

By "low frequencies" is meant

$$f < \frac{f_d}{(1 + L_b/ESL_d)^{1/2}} = f_L, \tag{3.12-5}$$

for which it has been shown in the discussion of Figure 3.3-3 that the input impedance of each of the K-DIP chains is approximately equivalent to a capacitor, $KC_d/[1 - (f/f_d)^2]^{1/2}$. Then, the equivalent circuit of a multiple-chain power bus is that in Figure 3.12-3.

In a way similar to that described for a single K-DIP chain in Section 3.3, low-frequency behavior of the power bus is different at different frequencies within the low-frequency range (3.12-5). The equivalent circuits given in Figure 3.12-4 are based on mathematical results derived in Appendix 1.

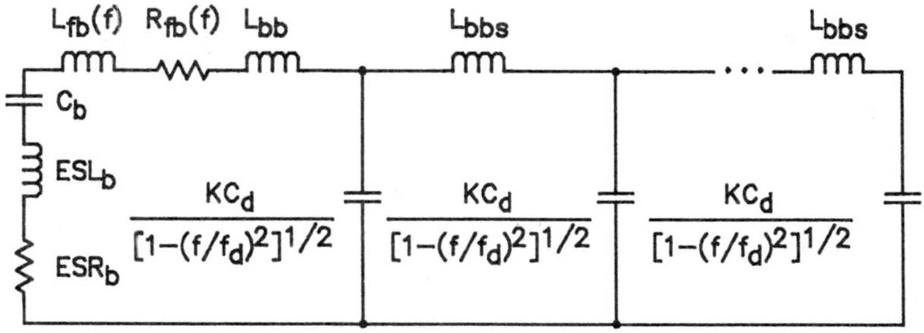


Figure 3.12-3: LOW-FREQUENCY EQUIVALENT CIRCUIT OF THE POWER BUS

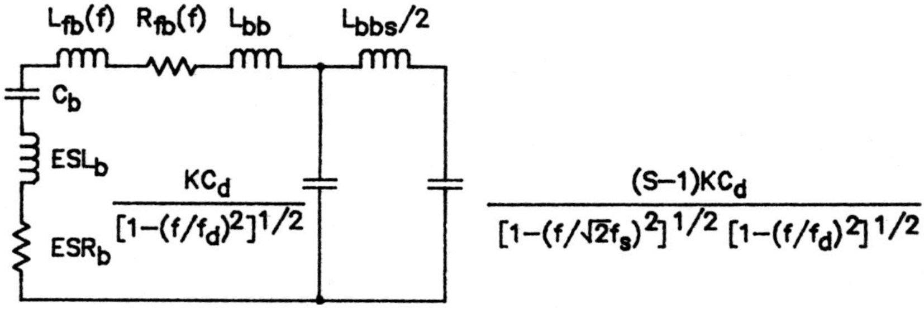


A. Equivalent Circuit for $0 < f < \{2\}^{1/2} f_s$ [Assuming $f_s < f_L$]

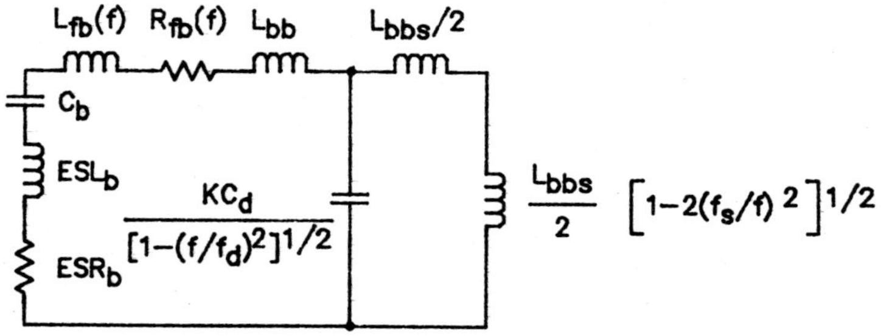


B. Equivalent Circuit for $\{2\}^{1/2} f_s < f < f_L$

Figure 3.12-4: THE CHARGING LOOP AT LOW FREQUENCIES

In the Figure,

$$f_s = \frac{2^{1/2}}{2\pi (KC_d L_{bbs})^{1/2}} \qquad (3.12\text{-}6)$$

In practice, element values are usually such that the possible ringing frequency falls within the range given in Figure 3.12-4A. In comparison to the inductance, $L_{fb}(f)$, of the ferrite bead at low frequencies, all other inductances are neglible. Then, the equivalent circuit in Figure 3.12-4A becomes a single loop with a right-hand-end capacitance of $KSC_d/[1 - (f/\{2\}^{1/2}f_s)^2]^{1/2}[1 - (f/f_d)^2]^{1/2}$. By analogy with the single-chain result (3.6-3),

$$D_{bs} \equiv \frac{4L_{bbe}(f)\,[1 - (f/f_d)^2]^{1/2}[1 - (f/\{2\}^{1/2}f_s)^2]^{1/2}}{(R_{fb}(f) + ESR_b)^2 KS(C_d + N[C_s + C_L])} < 1. \qquad (3.12\text{-}7)$$

$D_{bs}$ decreases (i.e., overdamping is progressively easier) for an increasing number, S, of rows of DIPs.

Implicit in the approximation (3.12-7) is the assumption that the value of the power entry capacitor is much greater than $KSC_d$. Selection of $C_b$ via

$$C_b \geq 10SKC_d \qquad [= 90SKN(C_s + C_L)] \qquad (3.12\text{-}8)$$

guarantees that the effective decoupling capacitance is decreased by no more than 9.1 %, and is recommended [the equality in brackets [] is from (3.5-3)]. An overly-large $C_b$ is, by virtue of increased physical size, likely to exhibit excessive $ESL_b$ and the undesirable emissions consequences associated with that condition.

**Discussion**

As a topological alternative to Figure 3.12-1, a "power-entry" capacitor and ferrite bead can be placed at the beginning of each of the S chains of DIPs. This nearly eliminates the (already minimal) aforementioned contributions to conducted noise from DIPs other than those in the first chain. In commercial practice, however, economy, and nearly-equivalent suppression, suggests use of a single power entry capacitor such as shown in Figure 3.12-1.

It is clear from the discussion of (3.12-2) that it is desirable to provide modest to moderate inductance between chains of DIPs so that the noise contributed at the power entry point by the second and subsequent chains of DIPs is, relative to that contributed by the first chain of DIPs, small. Again, this is a case where the traces should be arranged - or, for particularly demanding (usually military) applications, discrete inductors should be used - to provide modest to moderate inductance.

Oftentimes, one sees a requirement/recommendation for multiple (smaller-than-$C_b$) electrolytic capacitors at points intermediate to the power entry and decoupling capacitors (Figure 3.12-4). That requirement is removed - and economies improved - by employing a ferrite bead.

Of all difficulties encountered in practice as the number of chains of DIPs is increased, selection of a ferrite bead to provide the required $R_{fb}(f)$ **at the direct current level required by the PCB** causes the most concern. If the bead saturates due to excessive (for the bead) direct current, $R_{fb}(f)$ tends to zero and charging loop ringing occurs. In [3], it is shown that physically larger beads are necessary as current is increased. Alternatively, multiple smaller beads can be used.

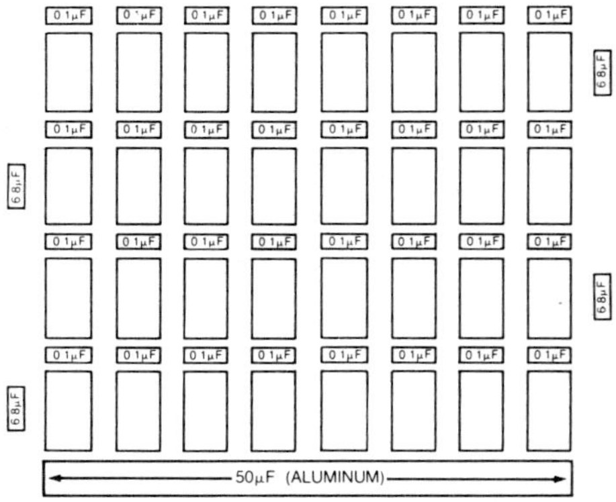


Figure 3.12-5: NOT RECOMMENDED

# 4.0 DESIGN RECOMMENDATIONS

## 4.1 PCB Layout

For convenient reference, Figure 3.2-1 - incorporating the layout which has been used as an example throughout this book - is repeated below.

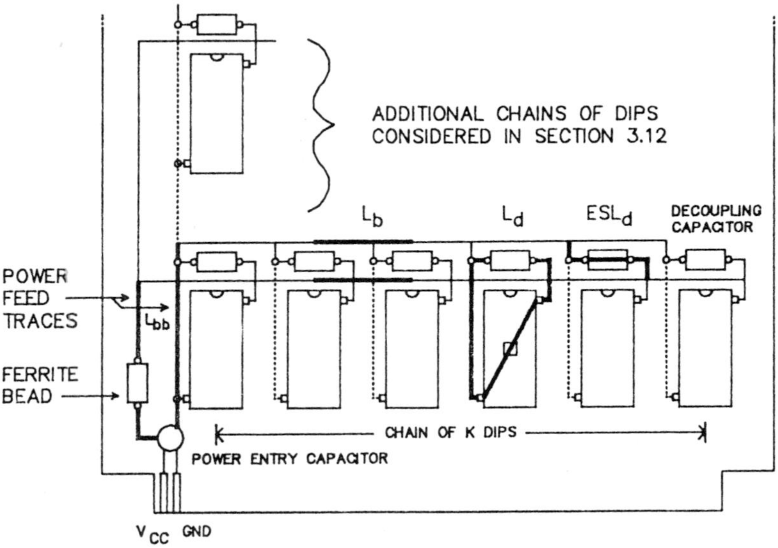


Figure 4.1-1: LAYOUT EXAMPLE

This example employs a "Good" DIP-local power bus layout, but there are worse and better layouts:

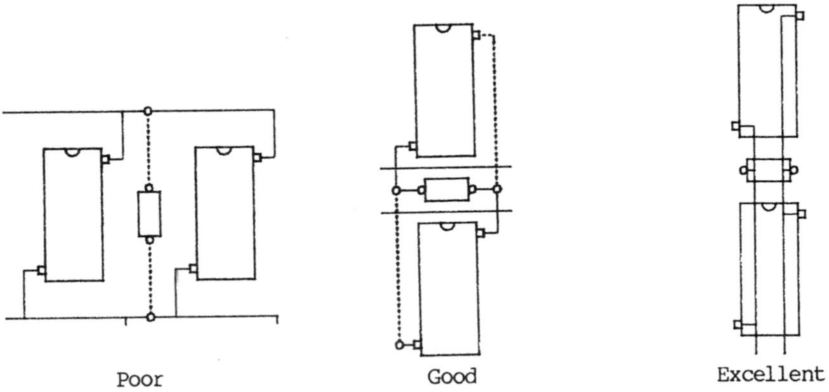


Figure 4.1-2: DIP-LOCAL POWER BUS LAYOUTS

The two-DIPs-per-decoupling-capacitor layouts shown in Figure 4.1-2 are recommended only when the analysis given in Section 3.11 indicates feasibility; wide DIP-to-bus traces and a small number of output gates help. In many practical cases, a decoupling capacitor will be required for each DIP.

Reasons for the ranking of the layouts in Figure 4.1-2 will become apparent in the discussions below, which are directed towards evaluating various layouts in terms of reducing on-board generated emissions and susceptibility to externally-generated radiated emissions (RFI).

**Radiated Emissions**

Clock lines, and not the power bus, are usually the worst direct radiators on a PCB. However, to insure that direct radiation is dominated by clock lines and to minimize both radiating and receiving loop areas of the power bus, the $V_{CC}$ and ground traces should be reasonably close together.

By virtue of a ground plane, multilayer PCBs accomplish that aim. However, a carefully laid out two-sided PCB will perform nearly as well. The dictum of [1, p. 4.2-7] is repeated here: **Do not incur the expense of multilayer PCBs just for EMI purposes.** Multilayer PCBs will "automatically" provide for radiation which is small in comparison to that of a poorly laid out two-sided PCB. However, the additional cost of a multilayer PCB is seldom justified by the minor radiated emissions reduction from that of a well laid out two-sided PCB.

There are, of course, reasons other than reduction of direct radiation to consider multilayer PCBs: reduction in ground noise, increased component density, and, for large production runs, improved economics relative to a larger number of two-sided PCBs.

**High-Frequency Conducted Emissions**

It is important to maintain the layout-dependent part of $ESL_d$ - traces connecting the decoupling capacitor across the power bus - at a minimum value in order to minimize high-frequency conducted emissions. Otherwise, both radiated and conducted emissions problems are possible via the various means outlined in Section 3.8.

Due to filtering effects provided by the power bus at high frequencies, **the first DIP in the chain of DIPs closest to the power entry point is the dominant source of noise.** Particular attention should be paid to minimizing $ESL_d$ for that chain of DIPs, and, within that chain, the DIP closest to the power entry point requires most attention.

The power bus should be layed out so as to accomodate minimization of that part of $ESL_d$ which appears across the power bus. Figure 4.1-2 exemplifies both good and bad layouts. The "Poor" layout requires relatively long decoupling capacitor leads/traces to connect that capacitor across the power bus. **Power bus layouts which provide for close spacing of $V_{cc}$ and ground traces and commensurate shortness of decoupling capacitor leads are good.**

The inductance $L_d$ stems in part from the PCB layout - traces used to connect the decoupling capacitor across the DIP - and in part from the inductance internal to the DIP. $L_d$ should be minimized to prevent conducted emissions which could otherwise be caused by ringing of the discharge loop.

The "Excellent" layout of Figure 4.1-2 is good from the standpoint of minimizing both layout-dependent $ESL_d$ and decoupling-capacitor-to-DIP inductance, $L_d$. The "Good" layout provides for low layout-dependent $ESL_d$, but $L_d$ is greater than that for the "Excellent" layout. The "Poor" layout has relatively large layout-dependent $ESL_d$ and $L_d$.

In effect, $L_d$ limits the maximum number of gates which can be effectively serviced by a decoupling capacitor without incurring high-frequency ringing of the discharge loop and the undesirable conducted emissions consequences associated with that ringing. However, $L_d$ is a tradeoff parameter: larger $L_d$ enhances the filtering of high-frequency conducted emissions, but too large a value causes ringing. As a design recommendation, the tradeoff should go in the direction of minimizing $L_d$, and good (low $ESL_d$) decoupling capacitors should be relied upon for high-frequency filtering. **Whenever a candidate layout suggests a tradeoff between minimization of $L_d$ and $ESL_d$, $ESL_d$ should take preference.**

**Other Power Bus Inductances**

When a ferrite bead is employed, other power bus inductances are not particularly critical. Within wide limits discussed below, those inductances can accommodate layout conveniences.

Use of a ferrite bead alleviates the need to minimize the inter-DIP power bus inductance, $L_b$, and the inductance of the power feed traces, $L_{bb}$, for the purpose of preventing charging loop ringing; the resistance of the bead accomplishes suppression of ringing even for quite large $L_{bb}$ and/or $L_b$. Thus, a ferrite bead permits $L_b$ and $L_{bb}$ to be large enough to be helpful in reducing conducted emissions. If a ferrite bead is used - as it should be - modest to moderate $L_{bb}$ is recommended.

In Figure 3.2-1, for example, it is not necessary (or, even, desirable) to take extraordinary measures to run the $L_{bb}$ traces immediately next to each other to reduce their inductance, as the larger inductance incurred with wider-spaced traces aids in filtering high-frequency noise which would otherwise be transmitted to the power-entry point. Wider spacing of the $L_{bb}$ traces does imply a more efficient power feed "loop antenna," suggesting possible increased radiated emissions. However, the (charging) current in the $L_{bb}$ traces is primarily at low frequencies (the clock and low harmonics thereof), where antenna efficiency is usually not large.

Current in the inter-DIP power bus traces, $L_b$, is at high frequencies and, therefore, increased radiation is a potential threat. However, due to the much larger area of the clock trace loops (which, of necessity, also carry high-frequency current), the radiated emissions spectrum is dominated by radiation from the clock traces. Then, $L_b$ can be modest to moderate - implying wider trace spacing and a somewhat larger loop area - without signifigantly affecting the overall radiated emissions spectrum characterizing the PCB. For this reason and because $L_b$ is useful in filtering high-frequency conducted noise, modest to moderate $L_b$ is recommended.

There are limits to artificially increasing trace spacing to obtain the larger inductances useful in filtering high-frequency conducted noise: given adequate PCB area and freedom from other constraints, the loop areas of both of the above otherwise non-critical loops should be kept small to minimize the EMI collection areas (i.e., antenna "pickup" efficiency) of the traces to outside signals. **Reducing loop areas always corresponds to a reduction of inductances and vice-versa.**

Consideration of special power bus products cited in Section 3.9 suggests that good conventional layout practices and good decoupling capacitors are to be preferred over those more expensive alternatives.

**Layout to Reduce Ground Noise**

In order to minimize ground noise, the highest-current DIPs should be as close as possible to the power entry point:

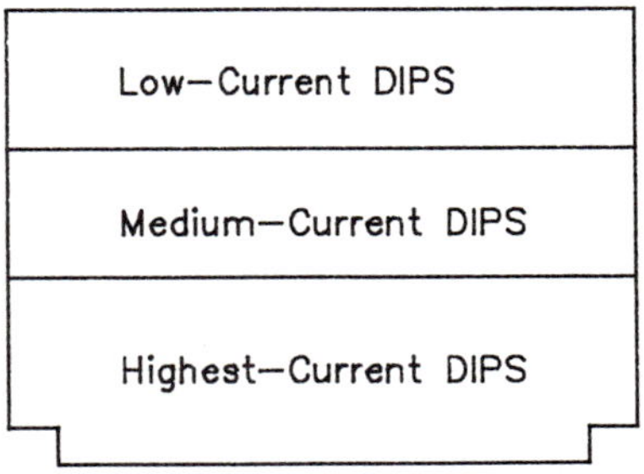


Figure 4.1-3: GROUP HIGH-CURRENT DIPS NEAR POWER ENTRY POINT

There are many other ways to reduce ground noise:

- Multilayer PCB with ground plane,
- Provision of a ground grid [10],
- Leave as much metal on the two-sided PCB as possible, and connect those "spare" metal islands together and to ground everywhere possible.

In usual two-sided PCB practice, the latter approach is the most practical.

**Summary of PCB Layout Recommendations**

Figure 4.1-4 provides details of an arrangement which experience has shown to perform well. That layout is amenable to the usual vertical trace runs on the component side of the PCB, and horizontal trace runs on the solder side of the PCB (or vice-versa).

Due to the increased antenna efficiency associated with longer PCB trace runs and the ground noise problems associated with those runs, multiple smaller PCBs are to be preferred over a lesser number of larger PCBs. It is recognized that, due to connectors, etc., multiple smaller PCBs are more expensive.

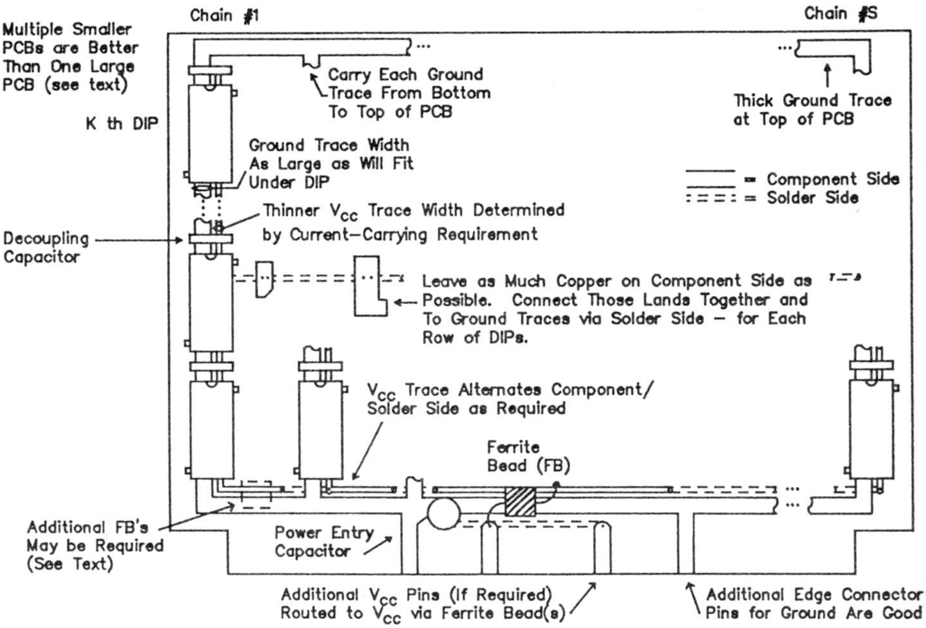


Figure 4.1-4: SUMMARY OF PCB LAYOUT RECOMMENDATIONS

In Figure 4.1-4, wide ground traces are used to minimize ground noise. Additionally, the horizontal ground runs connecting the spare lands together and to the vertical ground traces assist in minimizing ground noise.

The width(s) of the $V_{cc}$ traces are established by current-carrying requirements. The vertical ground traces should be as wide as possible, with limits set by the width of the DIP(s) and the following consideration: combined ground and $V_{cc}$ trace widths plus separation between those traces should not be so wide that solder wicking between either trace and the DIP pins occurs during wave soldering. Within this constraint, the ground trace should be as wide as possible.

Depending on the direct current required by the PCB, one large or multiple small ferrite beads may be required in order to accommodate that current without incurring saturation of the bead(s) [3].

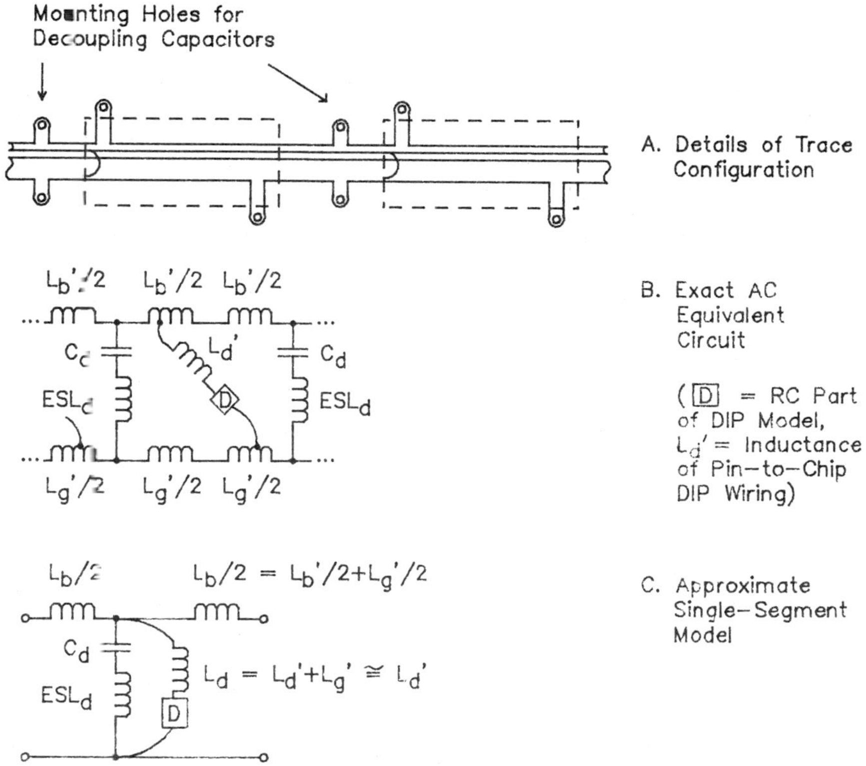


Figure 4.1-5: MODELING OF RECOMMENDED LAYOUT

Figure 4.1-5 provides a basis for modeling the layout shown in Figure 4.1-4. Relative to the layout shown in Figure 4.1-1, $L_d$ (= inductance incurred in connecting the semiconductor chip to the decoupling capacitor) is substantially decreased; in Figure 4.1-5, $L_d$ is very nearly the "irreducible" minimum of the inductance of the DIP-internal wire from the chip to the DIP pins. For a conventional 16-pin DIP, this is approximately 25.6 nH [see (A3-1) in Appendix 3].

For a ground trace much wider than the $V_{CC}$ trace, $L_g' \ll L_b'$ and the inductance $L_b$ is very nearly that of a trace next to a coplanar ground plane with a length equal to that between the mounting holes for the decoupling capacitors. From the upper graph on page A4-6 of Appendix 4, $L_b \cong 6$ nH/cm when the spacing between the ground and $V_{CC}$ traces is equal to the width of the $V_{CC}$ trace. Consistent with current-carrying and DC voltage drop requirements, the $V_{CC}$ trace should be as thin as possible so as to provide for moderate spacing between the ground and $V_{CC}$ traces. This increases $L_b$, an inductance which aids in filtering high-frequency noise.

## 4.2 Component Selection

The recommendations below, guided by the theoretical development of Section 3.0, reflect experience in the imperfect and cost-constrained environments in which most products are developed. When a component parameter has little effect, the specification is "loose."

The most important recommendation is to **thoroughly specify components.** One likely way to incur unpleasant surprises is to procure, say, decoupling capacitors, based soley on the lowest bid to an incomplete specification.

**Decoupling Capacitors**

• $C_d \geq 4.5 \times$ # Output Gates $\times (C_S + C_L)$ • $ESL_d < 20$ nH • $ESR_d \leq 0.5$ Ohms

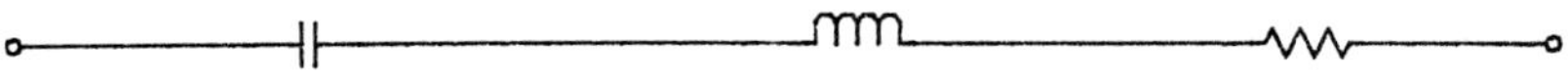

Leads/traces should add no more than 10 nH to $ESL_d$ (alternatively, total $ESL_d$ less than 30 nH). Double $C_d$ if two DIPs are served by one decoupling capacitor. Don't use more than two DIPs per capacitor.

Figure 4.2-1: SUMMARY OF DECOUPLING CAPACITOR RECOMMENDATIONS

Of all components to be specified, decoupling capacitors are most critical to minimization of high-frequency conducted emissions. However, a low-$ESL_d$ capacitor is only part of the problem. That capacitor must be connected to the power bus:

$$\text{Total } ESL_d = ESL_d \text{ of capacitor + lead and trace inductance.}$$

In addition to the above, the discussion below is based on the following observations:

- Generally, available decoupling capacitors suggest selection based on values which are an order of magnitude apart: 0.001 uF, 0.01 uF, and 0.1 uF are common values.

- The cost of 0.001 uF through 0.01 uF capacitors is about the same, but, for 0.02 through 0.1 uF - the latter being about the largest necessary when a ferrite is employed - the cost is somewhat larger.

Capacitance of Decoupling Capacitors

The number of output gates in a DIP which are going on at any given time is approximately half the number of total gates: N = M/2. To insure that the rise time is not increased, require

$$C_d \geq 9N(C_s + C_L) = 4.5M(C_s + FC_{in}). \qquad (4.2\text{-}1)$$

The latter term is determined from the worst-case output gate found in any of the DIPs in any single chain of DIPs:

N = M/2, M = Maximum number of output gates found in a DIP in the chain,

$C_s$ = Gate Ouput Capacitance (Table 3.2-1),

$C_L = FC_{in}$ = Worst-Case Fanout x Input Capacitance of Gate (Table 3.2-1).

Worst-case fanout is the maximum fanout found in a single chain of DIPs, with each output gate in the chain considered in determining that worst case.

On PCBs employing different logic families, it is wise to group like logic families in a given chain so that one DIP from a different logic family does not dominate the value of decoupling capacitors used in an entire chain. **Those technologies requiring larger values of decoupling capacitors should be grouped together in common chains so as to permit selection of lesser-valued (and less expensive) decoupling capacitors in other chains.**

For reasons discussed in depth in Section 3.11, the maximum number of DIPs per decoupling capacitor that should be considered is two - then only after careful layout and component selection to insure freedom from high-frequency ringing (see analysis of Section 3.11). If emissions are not a problem, a larger number of DIPs per decoupling capacitor can be considered.

Employing (4.2-1), the example carried throughout the present development leads to $C_d$ = 4.5 x 10 x (20 pF + 10 pF) = 1350 pF = 0.00135 uF; the next largest common value of decoupling capacitor is 0.01 uF, the "0.01" uF selected measured to be 0.007 uF - see Appendix 3.

Conducted emissions performance of the aforementioned example layout/circuit could be improved at lower frequencies by selecting a 0.2 uF decoupling capacitor with the same $ESL_d$. In terms of Figure 3.9-2, the null at 12.1 MHz is moved to

(4.2-2)

$$f_d = \frac{1}{2\pi(C_d ESL_d)^{1/2}} = \frac{1}{2\pi(2 \times 10^{-7} \times 24.8 \times 10^{-9})^{1/2}} = 2.26 \text{ MHz.}$$

Whether an increase to 0.2 uF is worth the added expense depends on the emissions and $ESL_d$ specifications. In military or FCC/VDE Class B designs concerned with relatively lower-frequency emissions, the added expense of a larger-valued decoupling capacitor may be justified if there are difficulties with, e.g., excess emissions in the vicinity of 2.26 MHz. However, it is clear from a comparison of Figures 3.9-5 and 3.9-6 that the emissions level at, e.g., 12.1 MHz will be increased (even allowing for the addition of a bead in the case shown in Figure 3.9-6).

Oftentimes, increasing the value of the capacitor also increses $ESL_d$, especially when the physical size of the capacitor must increase to provide the larger value. Any increase in $ESL_d$ is highly undesirable.

Intrinsic Inductance of Decoupling Capacitors

**The internal inductances of decoupling capacitors are at least as important as the values of those capacitors.** A large-$ESL_d$ capacitor adversely affects high-frequency conducted and radiated emissions and pulse integrity. Historically, ceramic capacitors have provided the least internal inductance consistent with low cost and low $ESR_d$.

Consistent with economic constraints, **select decoupling capacitors for minimum $ESL_d$.** Experience has shown that, with economical ceramic capacitors, internal $ESL_d$'s below 20 nH are both desirable and achievable; "average" values are 5 to 30 nH. Lead and trace inductance should add no more than 10 nH to that value, and the specification should include capacitor lead length.

Equivalent Series Resistance of Decoupling Capacitors

Experience has shown that $ESR_d$ should be maintained below 0.5 Ohms, which is neither difficult nor expensive. ESRd of practical capacitors varies with frequency. Data provided by manufacturers are usually for $ESR_d$ at the self-resonant frequency (only), where it is relatively easy to measure. These data are generally adequate for specification purposes.

**Power Entry Capacitors**

- $C_b \geq 10$ x Sum of Decoupling Capacitance on PCB (= 10 x Sum of $C_d$'s)

  - $ESL_b \leq 30$ nH, including leads/traces
  - $ESR_b \leq 1$ Ohm

Figure 4.2-2: SUMMARY OF POWER ENTRY CAPACITOR RECOMMENDATIONS

Capacitance of Power Entry Capacitors

From Section 3.9, the value of the power entry capacitor typically affects noise at less-than-clock frequencies.

The capacitance recommendation in the Figure is from (3.12-8). For the single-chain example (S = 1), that recommendation leads to $C_b > 0.42$ uF; a 4.7 uF capacitor was selected for a 6 x 6 matrix of DIPs requiring 6 x 0.42 = 2.52 uF.

A large (e.g., order of magnitude) increase in the value of the power entry capacitor beyond that required can lead to larger $ESL_b$ and cost.

Inductance of Power Entry Capacitor

From Section 3.9, inductance of the power entry capacitor affects clock and clock-harmonic frequencies. In particular, there is a strong dependence of (troublesome) VHF conducted noise on $ESL_b$.

Tantalum electrolytic and metallized polycarbonate capacitors provide internal inductances which are on the order of those for disk ceramic capacitors: 10 - 30 nH. They are also more expensive than aluminum electrolytic capacitors, for which the $ESL_b$ ranges to an order of magnitude or two higher (depending on the capacitor construction techniques used). That order of magnitude or two increase can translate to a 20 to 40 dB increase in conducted emissions at high frequencies, a very undesirable occurrence at those dangerously emissive (i.e., antenna-efficient) frequencies.

Purely as a consequence of emissions issues, **maintenance of $ESL_b$ below 10 - 30 nH, including leads and traces, is recommended.** The PCB should be layed out so that the inductance added by the leads of the capacitor and the traces connecting the capacitor across the power bus is minimal. Figure 3.2-1 provides an example.

Three and four-terminal capacitors reduce $ESL_b$ below that which is obtainable from conventional two-terminal capacitors, but they are more expensive.

The added expense of low-$ESL_b$ power-entry capacitors is warranted by emissions considerations, even for FCC Class A designs. Tantalum or metallized polycarbonate power entry capacitors are recommended. Aluminum types are not recommended due to their relatively large $ESL_b$.

<u>Equivalent Series Resistance of Power-Entry Capacitors</u>

When a ferrite bead is employed, the dominant effects of $ESR_b$ are at low frequencies, some of which typically include low clock harmonics.

At high frequencies where $ESR_b$ would otherwise be critical to suppression of noise, the resistance of the ferrite bead, $R_{fb}(f)$, is large. From the equations given in Section 3.9, it is apparent that this permits a larger value of $ESR_b$ than would otherwise be necessary for suppression of high-frequency conducted noise.

The 1-Ohm limit in the Figure above presumes the use of a ferrite bead. Maintenance of $ESR_b$ below that limit is neither difficult nor expensive.

**Ferrite Bead Selection**

- $L_{fb}(f_c)/[R_{fb}(f_c)]^2 < KSC_d/4$, $[f_c < f_d,\ f_c = \text{Clock Frequency}]$
- $30 \text{ Ohms } [@\ f > 5f_c] < R_{fb}(f) < 10 \text{ Ohms } [@\ f = f_c]$

Note: The above values should be maintained at the direct operating current.

Figure 4.2-3: SUMMARY OF FERRITE BEAD RECOMMENDATIONS

USE A FERRITE BEAD AT THE POWER ENTRY POINT. The advantages of doing so are

- Lower-value, physically smaller and less expensive decoupling capacitors,
- Higher-$ESL_b/ESR_b$ (less expensive) power entry capacitors,
- Easier PCB layout (more inductance is allowed), and
- Better conducted emissions suppression.

Bead selection criteria are implicit in the following equations:

| | |
|---|---|
| (3.6-4) | Tradeoff between value of decoupling capacitor and bead resistance necessary to achieve an overdamped charging loop. |
| (3.6-5) | Maximum value of bead resistance permitted under highly overdamped condtions. |
| (3.9-3) | Improvement in low-frequency conducted emissions due to bead resistance [set $R_{fb}(f) = 0$ to quantify performance without a bead]. |
| (3.9-6) | Improvement in high-frequency conducted emissions due to bead resistance. |

The recommendations in Figure 4.2-3 are good rules of thumb, but the details given in the above equations should be consulted.

Reference [3] addresses the problem of bead saturation with increased direct current, and provides measured characteristics of several beads of various physical sizes at 50 MHz. To maintain bead resistance, an increase in physical size is eventually required as the direct current is increased. Alternatively, multiple smaller beads - such as that used in the one-chain example - can be employed at the beginning of each chain of DIPs.

Some manufacturers provide data only on the magnitude of bead impedance, $Z_{fb}(f) = [R_{fb}^2(f) + \{2\pi f L_{fb}(f)\}^2]^{1/2}$, rather than on the separate resistive and inductive components. One (of many) inconveniences which result is an inability to determine the damping of the charging loop (3.6-3) or the value of the decoupling capacitor necessary to insure overdamping (3.6-4). Another lack in the specifications area is that some manufacturers do not provide data pertinent to bead saturation effects. **Select beads from manufacturers who specify and control inductance, resistance, and reduction of same due to direct current.**

## 4.3 Circuits Requiring Special Decoupling Care

Two circuit types require special care when selecting decoupling techniques:

- Clock generating/buffering DIPs and
- Input/Output (I/O) DIPs.

Maintenance of low emissions is the dominant reason for exercising special care when these DIPs are decoupled.

**Clock and Clock-Divider DIPs**

Clock circuitry, usually consisting of a crystal-controlled oscillator and buffer (sometimes with an interposed frequency divider), generates a relatively large amount of clock-related noise at the DIP(s) where those functions are exercised. If special care is not exercised, that noise is superimposed on the power bus, leading to the radiated and conducted emissions problems cited in Section 3.8.

It is recommended that a separate ferrite-bead Pi network be provided to reduce the transmission of clock-related noise to the $V_{CC}$ bus:

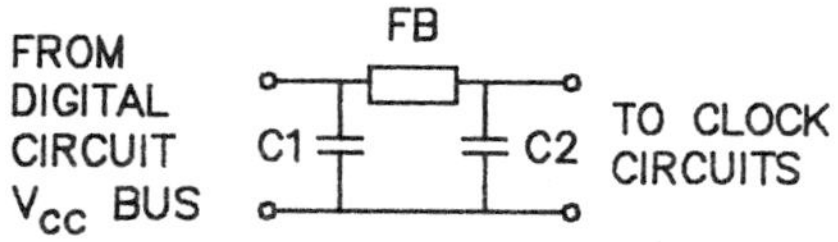


Figure 4.3-1: PI NETWORK FILTERING FOR CLOCK AND CLOCK DIVIDER DIPS

The value of C2 is chosen so as to provide the "normal" decoupling function required by the DIP(s) containing the clock functions. Depending on the physical distribution of those DIPs, more than one "C2" may be required, each of which is placed immediately adjacent to a relevant DIP.

The value of Cl is chosen to provide a low-impedance path to ground for all harmonics of the clock. As in the "normal" decoupling case, the ESL of Cl is, within wide limits, more important than its value. Typically, a 0.001 uF to 0.1 uF capacitor with ESL $<$ 20 nH is used. The value of Cl may be adjusted to "tune out" one or more particularly troublesome harmonics of the clock by selecting the resonant frequency of Cl, with lead/trace inductance considered, equal to the frequency of a troublesome harmonic or the center frequency of a group of troublesome harmonics.

It is recommended that clock DIPs be placed close to the power entry connector so as to minimize the effects of ground inductance in these relatively high-power DIPs.

**Input/Output DIPs**

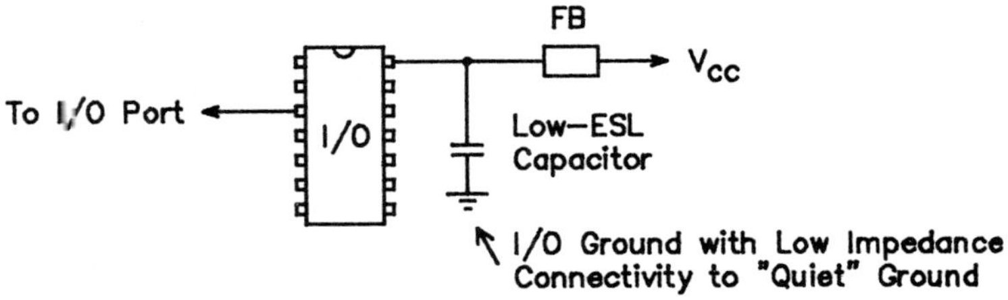


Figure 4.3-2: SPECIAL POWER BUS FILTERING FOR INPUT/OUTPUT DIPS

Contamination of the $V_{CC}$ bus serving the I/O DIPs can have disastrous consequences on radiated emissions (see discussion in Section 3.8). For that reason, it is recommended that strong consideration be given to creating a special separate $V_{CC}$ bus for those DIPs. That special bus is provided via

- A separate ferrite bead and power entry capacitor fed from the edge-connector side of the ferrite bead serving the digital logic DIPs, and

- A separate ground trace fed from the PCB edge connector, or, as is recommended in [10], attachment of the I/O DIP(s) ground via the chassis or other low impedance means to the single-point system ground.

# APPENDICES

# APPENDIX 1: DERIVATION OF POWER BUS MODELS

General Mathematical Background

In [9, p.186], it is shown that the ABCD matrix for a single segment

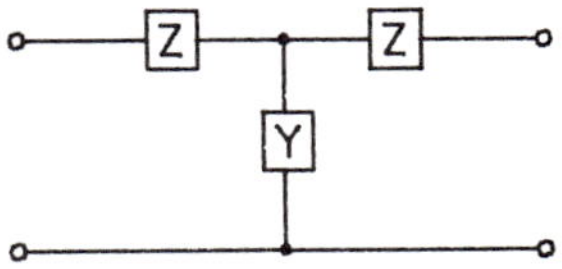


is

$$\begin{bmatrix} A & B \\ C & D \end{bmatrix} = \begin{bmatrix} 1 + YZ & 2Z + YZ^2 \\ Y & 1 + YZ \end{bmatrix} \qquad \text{(A1-1)}$$

The characteristic impedance of a cascade of such segments is

$$Z_o = (B/C)^{1/2} = [(2Z/Y)(1 + ZY/2]^{1/2} = Z(1 + 2/YZ)^{1/2} \qquad \text{(A1-2)}$$

If $|A| > 1$, the input impedance of a cascade of K of those segments, open-circuited at the far end, is [9, p. 188]

$$Z_{in} = Z_o \coth Ka \cong Z_o, \quad e^{Ka} >> e^{-Ka}, \qquad \text{(A1-3)}$$

where

$$a = \text{arccosh } A. \qquad \text{(A1-4)}$$

If, however, $|A| < 1$,

$$Z_{in} = - jZ_o \cot Ka, \qquad \text{(A1-5)}$$

where

$$a = \arccos A. \qquad \text{(A1-6)}$$

The above results are used three times in the text: for a chain of power bus segments, and for multiple rows of such chains at both high and low frequencies. Definition of high and low frequencies is given in (3.3-1), (3.3-2).

Chain of K Segments

The prototype segment for a chain of K DIP segments, each with a decoupling capacitor, is

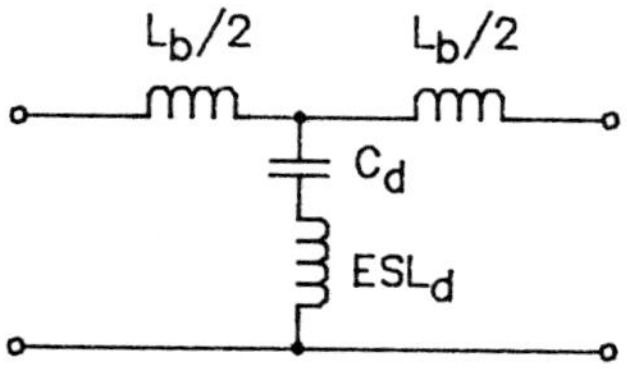


Then, from a comparison with the general YZ model above,

$$Y = \frac{j2\pi f C_d}{[1 - (f/f_d)^2]} \quad , \tag{A1-7}$$

$$Z = j2\pi f L_b/2 \ ,$$

where

$$f_d = \text{Self-resonant frequency of decoupling capacitor} = \frac{1}{2\pi(ESL_d C_d)^{1/2}} ,$$

Then,

$$A = [1 - \frac{(2\pi f)^2 C_d L_b/2}{[1 - (f/f_d)^2]}] \ , \tag{A1-8}$$

and, from (A1-2),

$$Z_o = j2\pi f \frac{L_b}{2} [1 - \frac{4(1 - \{f/f_d\}^2)}{(2\pi f)^2 L_b C_d}]^{1/2} . \tag{A1-9}$$

From (A1-8), it is easily shown that, for

$$f > \frac{f_d}{[1 + L_b/4ESL_d]^{1/2}} , \tag{A1-10}$$

$|A| > 1$. Then, for the condition $e^{Ka} >> e^{-Ka}$, it follows from (A1-3) that the (inductive) input impedance of a chain of K segments is, for the frequencies (A1-10), approximately (A1-9). That impedance divided by $j2\pi f$ is the L(f) given in the text. Similarly, $|A| < 1$ corresponds to low frequencies and the input impedance (Al-5) with $Z_o$ given in (Al-9). The low-frequency input impedance is very nearly the same as that of the physical models shown in the text.

S Rows of K-DIP Chains at High Frequencies

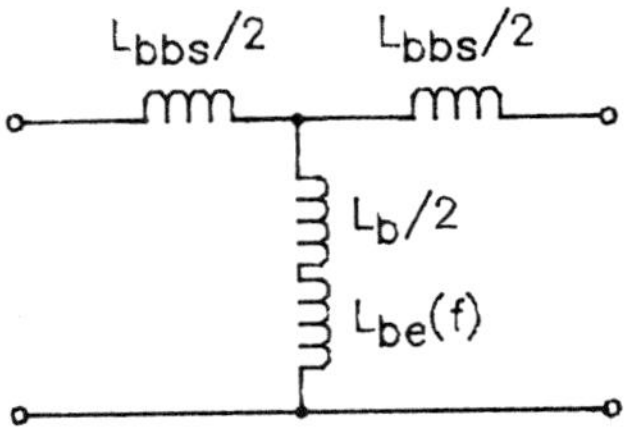


For this segment,

$$Z = j2\pi f L_{bbs}/2 \qquad \text{(Al-11)}$$

$$Y = \frac{1}{j2\pi f\{L_b/2 + L_{be}(f)\}} .$$

From (Al-1),

$$A = [1 + \frac{L_{bbs}}{L_b + 2L_{be}(f)}] \qquad \text{(Al-12)}$$

and, from (Al-3) with $e^{Sa} >> e^{-Sa}$,

$$Z_{in} \cong Z_o = j2\pi f(L_{bbs}/2)[1 + \{4L_{be}(f) + 2L_b\}/L_{bbs}]^{1/2}. \qquad \text{(Al-13)}$$

The value of the equivalent inductor given in the text is $L_S(f) \cong Z_{in}/j2\pi f$ plus $L_{bbs}/2$, the latter corresponding to the inductance preceding the first segment.

S Rows of K-DIP Chains at Low Frequencies

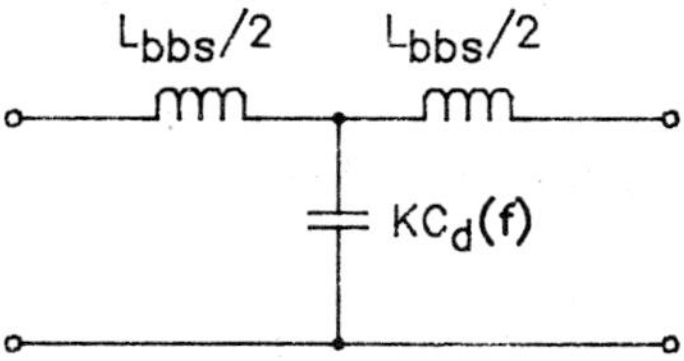


For this segment,

$$A = 1 - (f/f_s)^2, \quad \text{(A1-14)}$$

$$Z_o = j2\pi f \frac{L_{bbs}}{2} [1 - \frac{2}{(f/f_s)^2}]^{1/2},$$

$$= [L_{bbs}/KC_d(f)]^{1/2} [1 - (f/f_s)^2/2]^{1/2}$$

where

$$C_d(f) = C_d/[1 - (f/f_d)^2]^{1/2} \quad \text{(A1-15)}$$

$$f_s = \frac{1}{\pi[2L_{bbs}KC_d(f)]^{1/2}} .$$

For $0 \leq f < 2^{1/2}f_s$, $|A| < 1$, and, from (A1-5), the input impedance of (S - 1) segments is

$$Z_{in} = -j\, Z_o \cot [(S - 1)(\arccos A)] . \quad \text{(A1-16)}$$

For $f > 2^{1/2}f_s$, $|A| > 1$ and, from (A1-3), the input impedance of (S - 1) segments is

$$Z_{in} \cong Z_o, \qquad e^{(S - 1)a} \gg e^{-(S - 1)a}, \quad \text{(A1-17)}$$

where $a = \text{arccosh}\,|A|$. These input impedances correspond to those of the models given in the text.

# APPENDIX 2: AMPLITUDES OF SPIKES ACROSS DECOUPLING CAPACITORS

The purpose here is to determine the amplitudes of "spikes" across decoupling capacitors.

Straightforward analysis of the worst-case discharge loop in Figure 3.7-1 results in the following expression for the transform of voltage across the decoupling capacitor:

$$V_d(s) = \frac{(V_H - V_L)L_{be}(f)}{s[L_{be}(f) + L_d] + R_s/N + 1/s(C_s + C_L)}$$

Then, to obtain the amplitude of the voltage spike at t = 0,

$$\lim_{t \to 0} v_d(t) = \lim_{s \to \text{inf}} sV_d(s) = \frac{V_H - V_L}{1 + L_d/L_{be}(\text{inf})},$$

where "inf" denotes infinity. This is the result given in the text.

# APPENDIX 3: PARAMETERS FOR EXAMPLE

**Single-Segment Inductance Computations**

Figure A3-1 provides the detailed dimensions of the PCB layout shown in Figure 3.2-1. Figure A3-2 indicates the "local" element values to be determined.

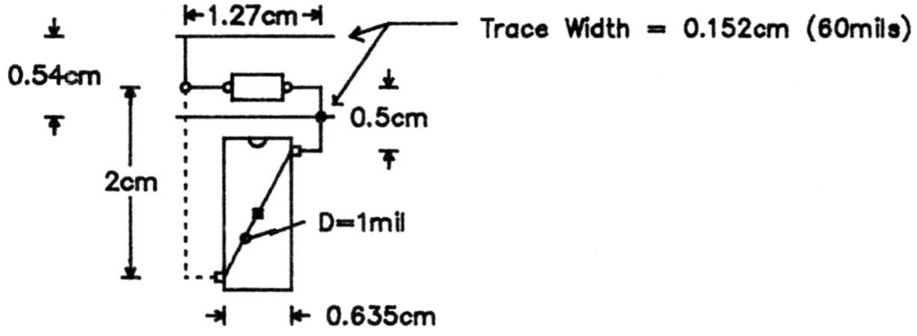


Figure A3-1: DETAILED DIMENSIONS OF SINGLE SEGMENT

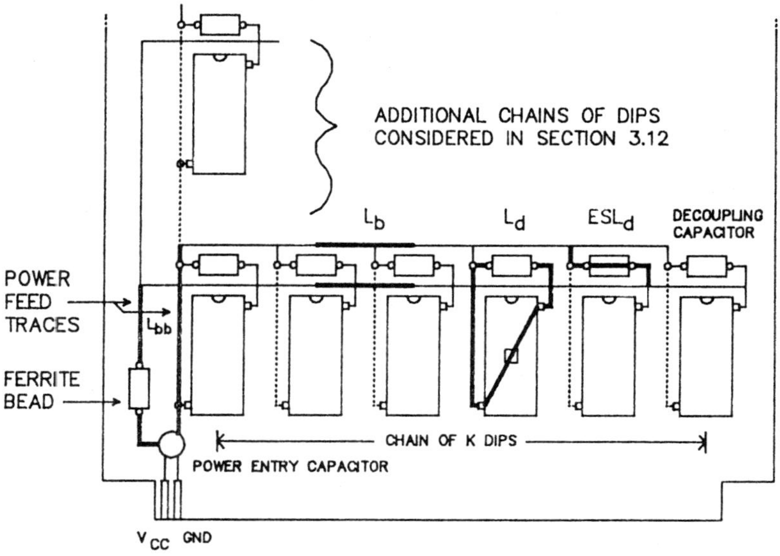


Traces: ——— = Component Side - - - - = Solder Side. FB = Ferrite Bead.

Figure A3-2: IDENTIFICATION OF ELEMENTS

The inductance of the 1 mil diameter wire internal to the DIP is approximated from [12, p.66]:

$$L \cong 2(\text{Length}) \ln [4(\text{Length})/\text{Diameter} - 0.75] \text{ nH}, \qquad \text{Length in cm}, \tag{A3-1}$$

$$= 25.6 \text{ nH}.$$

The inductance of the trace segments is approximated from [12, p.102]:

$$L \cong 2(\text{Length}) \ln \left[\frac{2(\text{Length})}{\text{Width}} + 0.5\right] \text{ nH}, \qquad \text{Length in cm}, \tag{A3-2}$$

$$= 27.9 \text{ nH},$$

so that

$$L_d \cong 25.6 + 27.9 = 53.5 \text{ nH}. \tag{A3-3}$$

The inductance values computed above assume that the wire/traces are layed out in a straight line (i.e., no mutual inductance). The effect of the (small) mutual inductance between the wires/traces is to slightly decrease the values from those determined above. The transmission-line technique used to determine the inductance of the horizontal portion of the power bus (see below) implicitly includes mutual inductance. That technique may be used to obtain the more accurate inductance values warranted when the spacing of the wires/traces is less than that characterizing the "local" area under consideration here.

The inductance of the power bus can be determined from the characteristic impedance of that bus. From Appendix 4, $Z_o$ = 315 Ohms. From (3.2-6),

$$L_b = 0.03333 \ (315) \times 1.27 = 13.3 \text{ nH}. \tag{A3-4}$$

Using (A3-1) with the diameter of the capacitor lead taken as that of # 22 gauge wire (0.0644 cm) and (A3-2), it is estimated that the inductance added to the self-inductance of the capacitor by the length of the capacitor's leads and the short segment of trace connecting the capacitor to the ground side of the bus traces is 6.8 nH:

$$ESL_d = ESL_{cap} + 6.8 \text{ nH} = 18 \text{ nH} + 6.8 \text{ nH} = 24.8 \text{ nH} \tag{A3-5}$$

Detailed characteristics of the capacitor are given in the Figure below:

Each lead = 0.2"

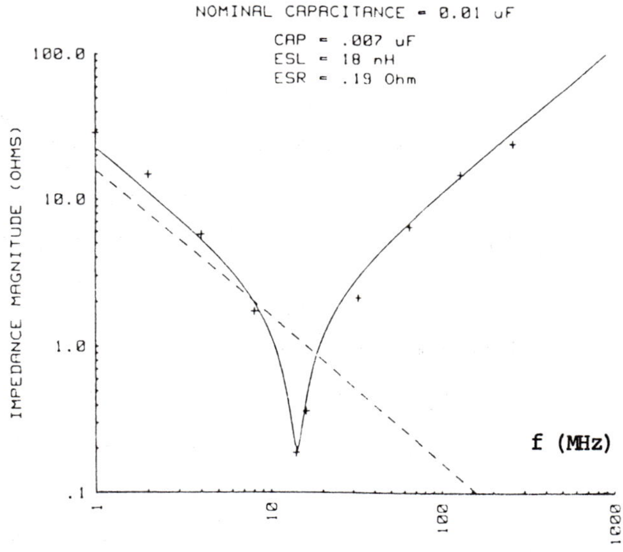


0.01 uF DISK CERAMIC CAPACITOR

**Inductance of Traces from Power Entry to First Segment**

In Figure 3.2-1, the length of the traces from the DIP side of the ferrite bead to the horizontal portion of the power bus is 2.2 cm; those traces are spaced 0.54 cm apart (same $Z_o$ as for horizontal portion of power bus). From (3.2-6),

$$L_{bb} = 0.03333\ (315) \times 2.2 = 23.1 \text{ nH.} \qquad \text{(A3-6)}$$

**Characteristics of Ferrite Bead**

The Figure on the next page indicates the characteristics of the Indiana-General AR-9701 bead used in developing the example. It should be noted that the low-current (Curve A) values have been used and that those values decrease with current (Curves B through D).

# AR-9701

## Mechanical Specifications

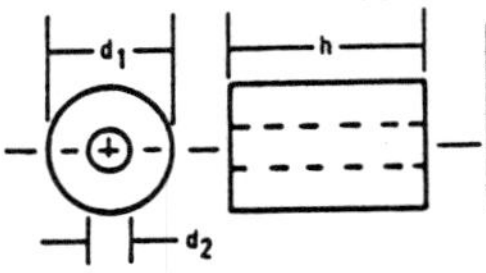


$d_1$ = 0.200 ±0.005 in.
$d_2$ = 0.050 ±0.005 in.
h = 0.160 ±0.010 in.
INS.

## Electrical Specifications

| D.C. Amp Turns | $\lvert Z_s \rvert / N^2$ 50 kHz | $\lvert Z_s \rvert / N^2$ 100 MHz |
|---|---|---|
| 0.0 Amp Turns | 2.58 Ωmin. | 42 Ωmin. |
| 0.7 Amp Turns | 0.129 Ωmin. | 20 Ωmin. |

Note: Measured with 1 turn < 5 mV

## Ampere Turn Parameters

(A) 0.0 NI (A turns)
(B) 0.17 NI (A turns)
(C) 0.37 NI (A turns)
(D) 0.70 NI (A turns)

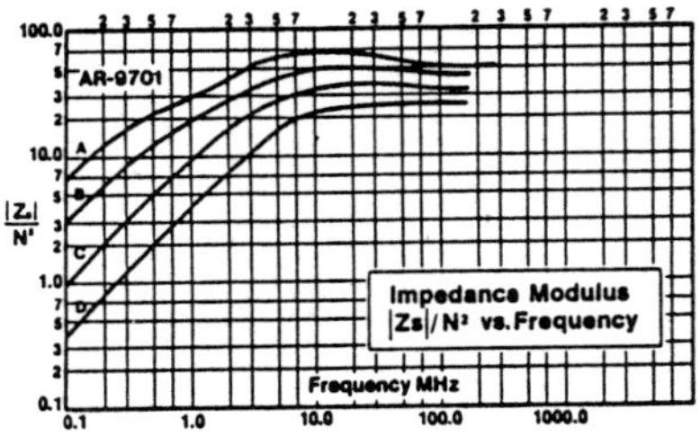


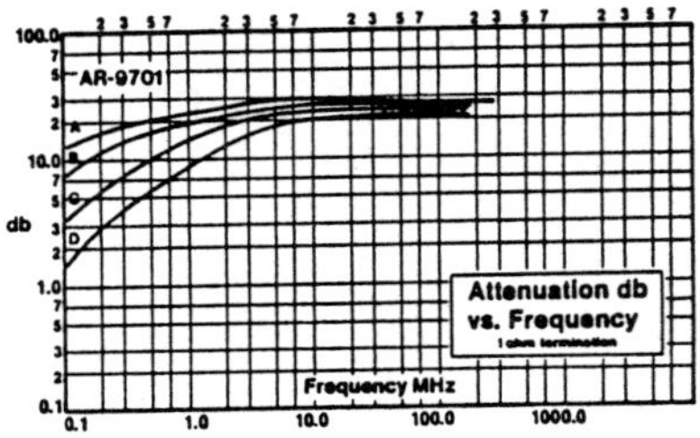


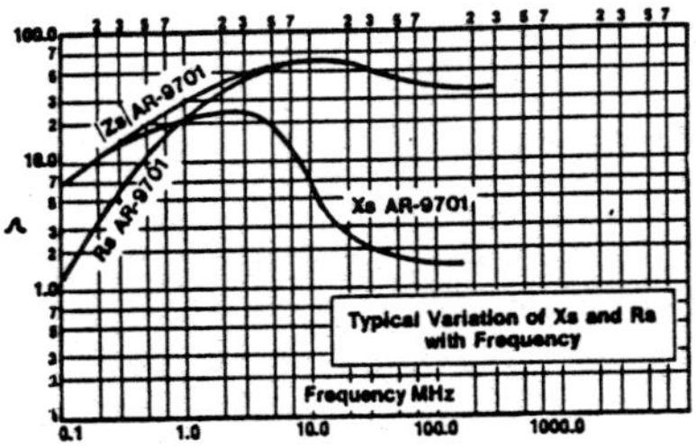


These data provided courtesy of Indiana General.

# APPENDIX 4: INDUCTANCE AND CAPACITANCE OF PCB TRACES

Figures providing inductance and capacitance for various PCB trace/ground plane configurations are presented at the end of the discussion below.

## Source and Accuracy

All characteristic impedance ($Z_o$) data presented here are from the excellent book by Hilberg translated from German by Claff [2]. Hilberg states that the accuracy of expressions for characteristic impedance is $2.4 \times 10^{-3}$ (0.24 %). An implicit assumption is that the thickness of traces is very much smaller than the width of traces, a condition which is met in most practical cases.

The reader interested in effects due to finite trace thickness and other details for the microstrip configuration (only) is referred to [13].

## Relationship Between Characteristic Impedance, Inductance and Capacitance

The characteristic impedance of a PCB trace configuration is defined as

$$Z_o = (L/C)^{1/2}, \qquad \text{(A4-1)}$$

where

L = Inductance per unit length,

C = Capacitance per unit length.

By methods to be outlined below, those elements are determinable from the exact relationships [2, p. 150]

$$L_o = \frac{0.100 \times Z_o}{3} = 0.03333 \times Z_o \quad \text{nH/cm}, \qquad \text{(A4-2)}$$

$$C_o = \frac{100}{3 \times Z_o} = 33.33/Z_o \quad \text{pF/cm},$$

The "o" subscript in $L_o$ and $C_o$ denotes inductance and capacitance (L and C) in air. $Z_o$ in (A4-2) is that determined for the traces in air.

Modifications for Non-Unity Relative Permittivity and Permeability

In PCB applications, relative permeability is unity ($u_r = 1$). Therefore,

$$L = L_o \, [u_r]^{1/2} = L_o \tag{A4-3}$$

i.e., **values for trace inductance determined from $Z_o$ for air apply to PCBs.**

For the purposes of determination of trace capacitance, let

$$e_r = \text{Relative permittivity of the PCB material} \quad (> 1), \tag{A4-4}$$

$$e_r' = \text{Effective relative permittivity for the PCB structure.}$$

Then,

$$C = [e_r']^{1/2} \, C_o. \tag{A4-5}$$

Part of the field is in air and part is in the PCB material; therefore,

$$1 < e_r' < e_r. \tag{A4-6}$$

In practice, $e_r$ ranges from approximately 3 to 7, and $e_r'$ ranges from 1.5 to 5. Therefore,

$$C_o < C < 1.22 \, C_o \text{ to } 2.24 \, C_o. \tag{A4-7}$$

Determination of accurate values for $e_r$ (and C) is a non-trivial electrostatics problem for all but the simplest structures. Biplanar traces (see Figures following) constitute a simple structure, and

$$e_r' \cong e_r, \tag{A4-8}$$

where the approximation arises due to fringing effects. Those effects are reduced (i.e., the accuracy of (A4-8) is improved) when the spacing between traces is small in comparison with the width of the traces.

To the author's knowledge, $e_r'$ for the practical case of coplanar traces on a PCB of finite thickness has not been analytically determined.

## Limits of Approximating a PCB Trace as a Pure Inductance

Limits are set by the length of the trace runs in relation to a wavelength at the highest frequency of interest, about 300 MHz for digital logic circuitry. Provided that a trace run is much less than the 4.4" to 8" (depending on $e_r$) given in Section 3.2, the impedance of a PCB trace run (or that of a trace/ground plane configuration) is

$$X_L = j2\pi f\,L, \qquad \text{(A4-9)}$$

where L is determined from (A4-3). Should a trace run exceed that length, the capacitance of the run becomes significant in determining the trace impedance; i.e., the trace run becomes a "transmission line." The reader is referred to [1] for a full discussion of effects.

## Equations Used to Determine Characteristic Impedance

For the purposes of precision, equations used to plot the curves are given below. Let

$Z_{oo} = Z_c/Z_{of}$, characteristic impedance of structure normalized to that of free space ($Z_{of}$ = 377 Ohms). $Z_{oo}$ is dimensionless.

$Z_{oo}$ below is for conductors in an air ($e_r$ = 1) medium. Relative to the expressions in [2], the coordinate system is transformed to maximize convenience; T is the thickness of a PCB, S is the spacing between the nearest edges (not centers) of coplanar conductors, and W is trace width.

**Trace Next to Coplanar Ground Plane** [2, p. 129]

$$Z_{oo} = \frac{1}{2\pi} \ln \left[2\,\frac{(W/S + 1)^{1/4} + 1}{(W/S + 1)^{1/4} - 1}\right], \qquad 0 < W/S < 32.5$$

$$= \frac{\pi}{8 \ln [2(W/S + 1)^{1/4}]}, \qquad 32.5 < W/S < \text{Infinity}$$

**Coplanar Traces** [2, p. 122]

$$Z_{oo} = \frac{1}{\pi} \ln \left[ 2 \, \frac{(2W/S + 1)^{1/2} + 1}{(2W/S + 1)^{1/2} - 1} \right], \qquad 0 < W/S < 2.40$$

$$= \frac{\pi}{4 \ln [2(2W/S + 1)^{1/2}]}, \qquad 2.40 < W/S < \text{Infinity}$$

**Biplanar Traces** [2, p. 140]

$$W/T = \frac{4}{e^{\pi Z_{oo}}}, \qquad 0 < W/T < 0.465$$

$$= \frac{\left[ \frac{\pi}{Z_{oo}} - \ln 2 - \ln \left( \frac{\pi}{Z_{oo}} - \ln 2 \right) - 1 \right]}{\pi}, \qquad 0.465 < W/T < \text{Infinity}$$

**Trace Over a Ground Plane (Microstrip)** [2, p. 140]

$$W/T = \frac{8}{e^{2\pi Z_{oo}}}, \qquad 0 < W/T < 0.91$$

$$= \frac{2\left[ \frac{\pi}{2Z_{oo}} - \ln 2 - \ln \left( \frac{\pi}{2Z_{oo}} - \ln 2 \right) - 1 \right]}{\pi}, \qquad 0.91 < W/T < \text{Infinity}$$

**Trace Between Two Ground Planes (Stripline)** [2, p. 131]

$$Z_{oo} = \frac{1}{2\pi} \ln [2 \coth (\pi W/4T)], \qquad 0 < W/T < 0.56$$

$$= \frac{\pi}{8 \ln [2e^{\pi W/2T}]}, \qquad 0.56 < W/T < \text{Infinity}$$

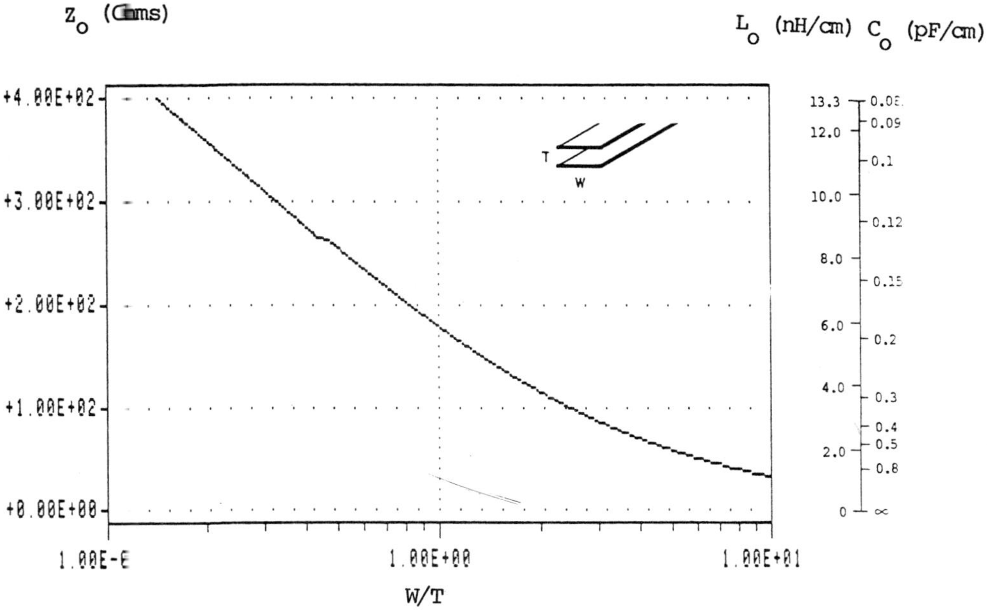


BIPLANAR TRACES

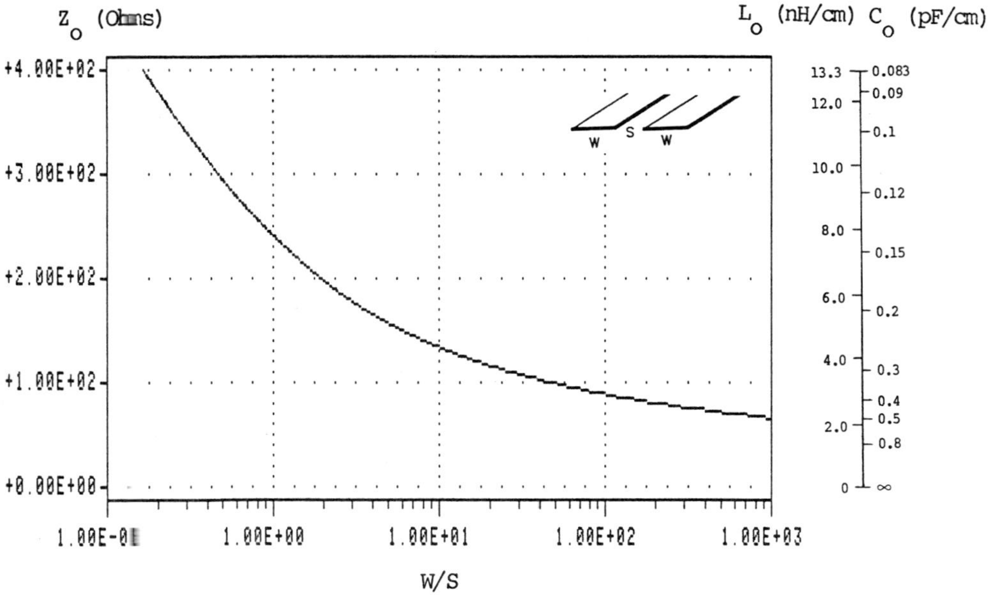


COPLANAR TRACES

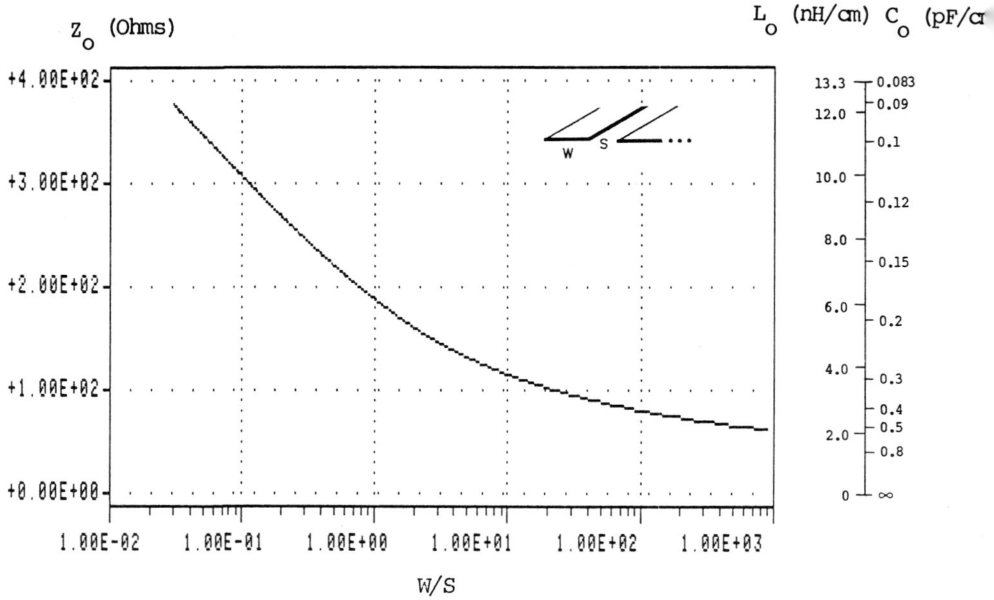


TRACE NEXT TO A COPLANAR GROUND PLANE

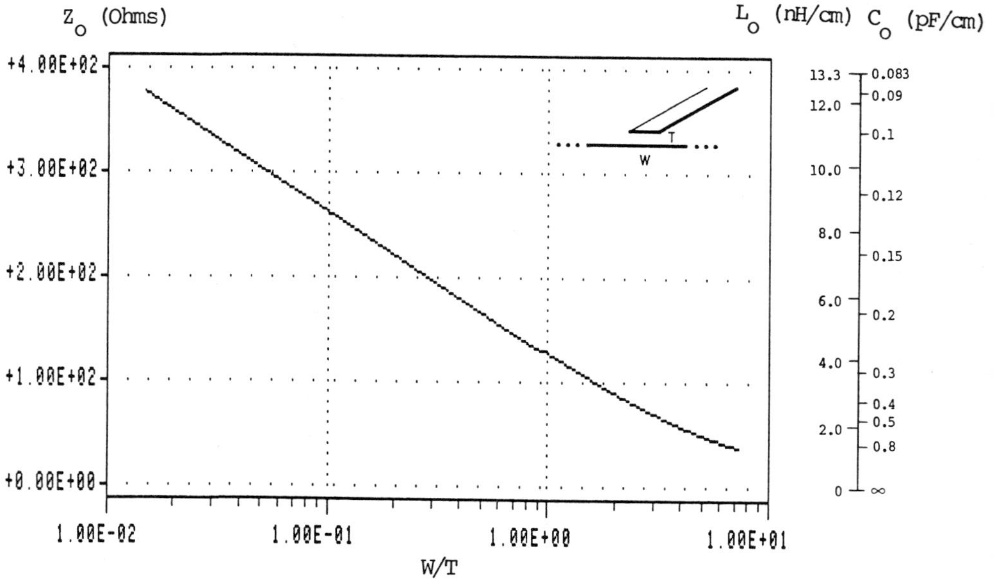


TRACE OVER A GROUND PLANE (MICROSTRIP)

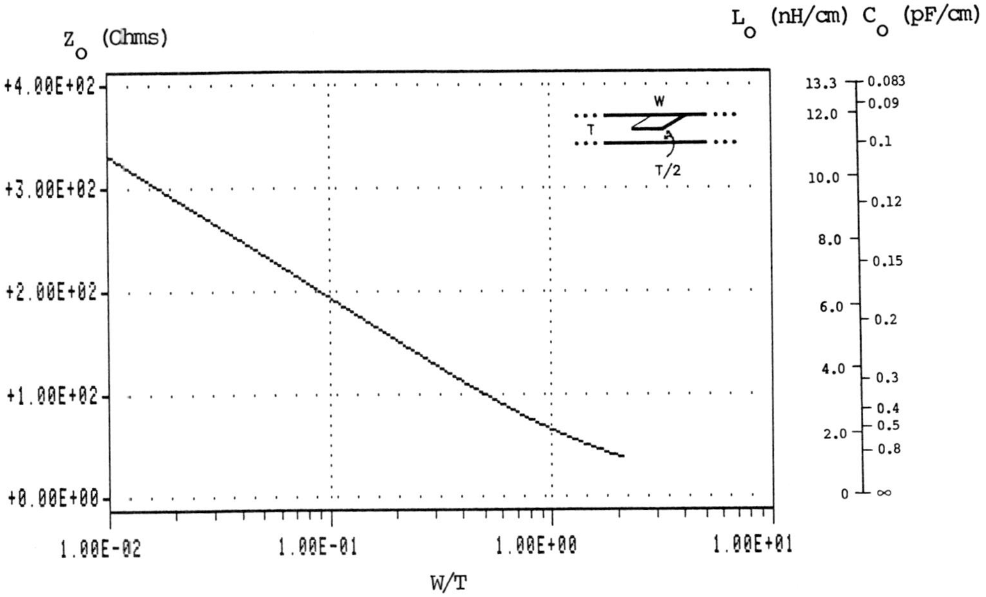


TRACE BETWEEN TWO GROUND PLANES (STRIPLINE)

# BIBLIOGRAPHY

[1] R. K. Keenan, **Digital Design for Interference Specifications,** Pinellas Park, Florida: TKC, 1983.

[2] W. Hilberg, **Electrical Characteristics of Transmission Lines,** Dedham, Massachusetts: Artech House, 1979.

[3] R. K. Keenan, "FCC/VDE Noise Specifications: Application of Ferrite Beads to Decoupling Printed-Circuit Boards from Backplanes", TKC Report dated 4 April, 1984.

[4] Motorola Technical Information Center, **CMOS Data Book,** Austin, Texas: Motorola, 1978.

[5] Motorola Technical Information Center, **MECL Data Book,** Phoenix, Arizona: Motorola, 1982.

[6] National Semiconductor Corporation, **High Speed Micro CMOS™ Logic Family Data Book,** Santa Clara, California: National Semiconductor, 1983.

[7] Texas Instruments Incorporated Semiconductor Group, **The TTL Data Book for Design Engineers,** Dallas, Texas: Texas Instruments Incorporated, 1983.

[8] J. A. Aseltine, **Transform Method in Linear System Analysis,** New York: McGraw-Hill, 1958.

[9] L. A. Pipes, **Applied Mathematics for Engineers and Physicists,** New York: McGraw-Hill, 1958.

[10] H. W. Ott, "Controlling EMI by Proper Printed Wiring Board Layout", **Proc. 1985 Zurich Conference on EMC.**

[11] J. W. Parker, Workshop at 1983 IEEE EMC Symposium, Washington, DC, and "A Useful Model for Evaluating Cable Radiation", **Proc. 1984 IEEE Conference on EMC,** pp. 371-374.

[12] B. E. Keiser, **Principles of Electromagnetic Compatibility,** Dedham, Massachusetts: Artech House, 1981.

[13] T. C. Edwards, **Foundations for Microstrip Circuit Design,** New York: Wiley, 1981.

# INDEX

## ABOUT THE AUTHOR

R. Kenneth Keenan was born in Pueblo, Colorado, in 1938. He holds the B.Sc.Engr. (with honors) from UCLA, the MSEE from Cal Tech, and the Ph.D. (EE) from Monash University, Melbourne, Australia. He is a member of IEEE, Tau Beta Pi, and Sigma Xi, and is listed in American Men of Science and Who's Who in Finance and Industry. He has published many papers in the areas of circuit theory, communications, and signal processing and is author of the standard technical reference book for FCC/VDE and MIL-STD emissions suppression design, **Digital Design for Interference Specifications** (TKC: 1983). After twenty years in industry in capacities ranging from Technician to Director of Engineering, he founded TKC in 1979. He is Vice President, Engineering, at TKC.

TKC is a consulting firm specializing in Electromagnetic Compatibility (EMC), including interference and transient upset mitigation techniques and product design to meet commercial and military specifications.